New Jersey Driver Tests:

700+ Questions, All-Inclusive Driver Practice Handbook to Quickly achieve your Driver's License or Learner's Permit

(Cheat Sheets + Digital Flashcards + Mobile App)

Written by Stanley Vast
Contributor - *Vast Pass Driver's Training*

© Copyright 2021 - All rights reserved.
Stanley Vast - Vast Pass Driver's Training

The content contained within this book may not be reproduced, duplicated or transmitted without direct written permission from the author or the publisher.

Under no circumstances will any blame or legal responsibility be held against the publisher, or author, for any damages, reparation, or monetary loss due to the information contained within this book. Either directly or indirectly. You are responsible for your own choices, actions, and results.

Legal Notice:

This book is copyright protected. This book is only for personal use. You cannot amend, distribute, sell, use, quote or paraphrase any part, or the content within this book, without the consent of the author or publisher.

Disclaimer Notice:

Please note the information contained within this document is for educational and entertainment purposes only. All effort has been executed to present accurate, up to date, and reliable, complete information. No warranties of any kind are declared or implied. Readers acknowledge that the author is not engaging in the rendering of legal, financial, medical or professional advice. The content within this book has been derived from various sources. Please consult a licensed professional and check with your local MVC for their manual as the techniques outlined in this book may have changed.

By reading this document, the reader agrees that under no circumstances is the author responsible for any losses, direct or indirect, which are incurred as a result of the use of the information contained within this document, including, but not limited to, — errors, omissions, or inaccuracies.

Table of Contents

Introduction 1

Chapter 1: Driving Under the Age of 21 3

Chapter 2: Understanding the Rules of the Road 12

Chapter 3: Recognizing Road Signs 36

Chapter 4: Traffic Lights & Signals Awareness 43

Chapter 5: Turns & Intersection Maneuvers 52

Chapter 6: True or False Trivia 66

Chapter 7: Driving Under the Influence 73

Chapter 8: Developing Safe Driving Habits 82

Chapter 9: Defensive Driving Techniques 94

Chapter 10: The Super Bowl 110

Chapter 11: Bonus Cheat Sheet I 132

Chapter 12: Bonus Cheat Sheet II 145

Chapter 13: Preparing for Your Exam 157

Chapter 14: Applying for Your Driver's License or Permit . 159

Conclusion 162

Would You Leave Us A Review? 164

Resources Page 167

Subscribe to Get Your
Private Mobile App
➕
Digital Flashcards

Go to NJ.Vast-Pass.com to get your Private Mobile App & Digital Flashcards!

Introduction

For over 15 years, Stanley Vast, a Driver's Health and Safety Consultant and Vast Pass Training Instructor has observed the challenges many individuals are facing every day when preparing for their driver's license or learner's permit. Stan created Vast Pass Driver's Training to assist teens, young adults and aspiring driver's in achieving their goals and obtaining the practice and education required to successfully pass their Department of Motor Vehicle exams.

Vast Pass Driver's Training provides a more in-depth and laser focused understanding of the problems many teens and young adults face when taking their driver's license and learner's permit exam. Stan created this course to help bridge this educational gap without exorbitant unreasonable fees. Since its launch, more than 2300 students have gone on to achieve their goal of becoming responsible and effective drivers. Armed with this knowledge you too can achieve your goals through this course, you can obtain the same success as many others have before you.

So, if you are nervous and feeling unprepared. No worries. You will acquire all the essential foundations and skills needed to become successful both on the road and in preparation for your exam.

In this book, we will uncover how you can easily prepare for your driver's license or learner's permit exam without any prior knowledge. You will also gain real-world insights on how to effectively handle road conditions and develop healthy driving habits.

The chapters in this book will discuss each of the following:

- Driving under the age of 21
- Road rules, habits, behaviors and techniques
- Road signs, traffic lights, and signals
- Defensive driving strategies
- Driving under the influence
- Strategies for success on your exam
- Processes for obtaining your driver's license and learner's permit

In addition to these topics, you will also be provided a couple of additional practice questions, cheat sheets, and knowledge checks to help you practice and prepare for your exam. Without further ado, let's begin.

Chapter 1
Driving Under the Age of 21

In this Chapter, we'll be discussing Driving Under the Age of 21 as it pertains to the law. With so many different rules and regulations to be aware of when you hit the road, you must familiarize yourself with each of them. In this chapter, you'll learn about the effects of alcohol on your ability to drive and how to handle situations such as crashes and interactions with law enforcement officers. In-depth information about breath, blood, and urine tests will be presented as well as material regarding blood alcohol content (BAC) and driving under the influence. Familiarizing yourself with each of these will help ensure you know exactly what is required of you to stay as safe as possible while on the road. There will be 12 multiple-choice questions for you to test yourself and see what you've learned.

From this chapter, you will develop a strong understanding of the potential consequences of failing to adhere to the standards set up by the law. You should also aim to understand the risks involved when getting behind the wheel and the factors that may make you more prone to an accident. Your ability to understand the techniques that promote safe driving will greatly influence your success as a new driver and ensure the safety of both yourself and those around you.

Let's begin:

1. If you are driving and a police officer stops you, you should:

   A. Take off your seatbelt and lower the car's window to address their concerns.
   B. Hurry and ready your paperwork such as license and registration before the concerned officer reaches your vehicle.
   C. Keep your calm and stay within the vehicle with your hands firm on the steering wheel as you wait for the officer to approach you.
   D. Get out and walk towards the police car and seek the officer's concerns.

Correct Answer is c

(When you are driving and a police officer stops you, you should keep calm. It is important that you keep your hands at the steering wheel at all times. Make sure not to make any sudden moves and wait until the concerned officer asks you to show your documents)

2. **If you are under the age of sixteen and use false identification to purchase alcohol, you are likely to:**

    A. Receive a driving suspension commencing on your sixteenth birthday.
    B. Only be allowed to take the driver's exam after you are twenty-one years old.
    C. Receive a driving suspension commencing on your twenty-first birthday.
    D. Be forced to take an alcohol safety education class.

    Correct Answer is a

    (Although you may not be driving, you could receive a suspension from driving if you are caught doing any of these three things; possessing alcohol, faking your age to get alcohol, or carry false identification)

3. **If you are caught driving under the influence of alcohol, you could be charged with:**

    A. A driver's license suspension for up to 3 years.|
    B. A hundred dollar fine or more.
    C. A compulsory attendance to Alcohol Highway Safety School.
    D. A twelve-hour jail sentence or more.

Correct Answer is c

(Anyone caught driving under the influence will be asked to attend a Highway Safety School for their first or second offenses. This is applicable for any level of impairment)

4. If you get involved in a car crash while driving, it is imperative that you must:

- A. Ensure that the injured person reaches the nearest hospital.
- B. Continue to drive to your destination, and later file a crash report.
- C. Stop your vehicle and help anyone who might be injured. Then you must report the crash to the police and exchange information. You must also notify your insurance company.
- D. Stop the car for a bit and check if there have been any damages to it.

Correct Answer is c

(If you end up getting involved in a car crash, you must stop at the scene. Assist and help anyone who might be injured. Exchange your personal information with others who have been involved in the crash. Secondly, you must also report the crash to the police as soon as possible and contact your insurance company)

5. Parental agreement to conduct breath, blood and urine tests is:

A. No parental consent is required to conduct such tests.
B. Yes, consent is required though only from one of the parents.
C. Yes, such consent is necessary for individuals under sixteen years of age.
D. Yes, such consent is required from both parents of the ward.

Correct Answer is a

(If you are arrested when under the age of twenty-one, you can be detained for blood, breath, or urine testing. The authorities require no parental consent for such tests)

6. If you happen to be driving under the influence of alcohol and refuse to give a blood test, you are likely to receive:

A. Treatment of drug counseling for 180 days.
B. A day-long sentence in jail.
C. Suspension of your driver's license.
D. Three hundred dollars fine due in 30 days.

Correct Answer is c

(If you refuse to take any of the chemical tests for breath, urine, or blood in the instance that a police officer arrests you for driving under substance influence, your driving privilege will automatically be suspended)

7. Anyone who is under the age of twenty-one is not legally permitted to _____ alcohol.

    A. Serve any.

    B. Wear clothes advertising any.

    C. Be near any.

    D. Transport any.

Correct Answer is d

(It is unlawful for anyone under the age of twenty-one to buy or consume alcohol. They are also now allowed to have it in their possession or transport it in a vehicle while they are driving)

8. Anyone who is under the age of twenty-one is not legally permitted to ____ alcohol.

   A. Wear clothes advertising any.
   B. Be near any.
   C. Consume any.
   D. Serve any.

   Correct Answer is c

   (It is unlawful for anyone under the age of twenty-one to buy or consume alcohol. They are also now allowed to have it in their possession or transport it in a vehicle while they are driving)

9. Anyone who is under the age of twenty-one is not legally permitted to ____ alcohol.

   A. Wear clothes advertising any.
   B. Possess any.
   C. Serve any.
   D. Be near any.

Correct Answer is b

(It is unlawful for anyone under the age of twenty-one to buy or consume alcohol. They are also now allowed to have it in their possession or transport it in a vehicle while they are driving)

10. **When the zero-tolerance law was introduced, the acceptable blood alcohol content (bac) was reduced from 0.08% to a ___ % for individuals under the age of twenty-one who were driving under the influence.**

    A. .02%
    B. .05%
    C. .07%
    D. .00%

Correct Answer is a

(It is possible for a young driver to be charged for driving under influence even if they happen to have consumed only a small glass of wine with dinner)

11. If you are caught to be driving under the influence, and a police officer requires you to take a breath, urine, or blood test, you:

   A. You can test at your own leisure.

   B. Must also sign a consent form with the test.

   C. You don't have to test since you're under 21.

   D. You must take the test, or your license could be suspended.

   Correct Answer is d

   (Your driver's license will automatically be suspended if you are arrested by the police and refuse to take a breath, blood, or urine test)

12. If you are a driver under age twenty-one, and are convicted of having a false id, you could be required to pay a five hundred dollar fine even if:

   A. Your blood alcohol content is found to be slightly greater than 0.02%.

   B. You are not the driver or passenger.

   C. You were sleeping.

   D. Your blood alcohol content is 0.01%.

   Correct Answer is b

   (Your will be fined even if you are not driving)

Chapter 2
Understanding the Rules of the Road

Possessing a driver's license comes with a lot of responsibility as well as a certain set of rules that must be followed every time you get behind the wheel. Chapter 2 will introduce the regulations that have been put in place to make the road as safe as possible for all drivers. We'll be walking through speed limits, traffic patterns, signaling, dangerous road conditions, and much more. While it's essential that you pay close attention to what your own vehicle is doing, safe driving also means you need to keep an eye on the traffic and others around you. Driving conditions can change rapidly and it's important to be prepared for anything that may come your way. Whether it's a bicyclist, heavy rainfall, or a car that has suddenly crossed your path, you'll find that the road will demand your full attention at all times.

The best way to prepare yourself to be a great driver is to center your driving habits around rule-following from the start. You can use the 39 questions in this section to test your Knowledge and Understanding of the Rules of the Road. You'll be able to get an idea of the areas you've already mastered as well as the ones that need more review.

Let's begin:

1. If the back of your vehicle starts to skid to the left, you should:

    A. Hold the steering wheel tight and steer left.

    B. Immediately slam on your brakes.

    C. Get in the fast lane and accelerate.

    D. Stop your vehicle.

Correct Answer is a

(Tightly hold the steering wheel if the rear of your vehicle starts to swerve. Begin steering to the left)

2. You may only cross a double yellow line when passing another vehicle if the yellow line is:

    A. On the other side of the road and line is dotted.

    B. On your side of the road and line is broken.

    C. On the other side of the road and line is dotted.

    D. On your side of the road and line is solid.

New Jersey Driver's Practice Tests

Correct Answer is b

(A double yellow line in the center of the road indicates that you may pass only if a broken line is next to your lane)

3. **When you are driving on a highway and need to know the distance to the next exit, what color of a sign should you be looking for?**

    A. Black color lettering on a yellow board.

    B. White color lettering on a black board.

    C. White color lettering on a red board.

    D. White color lettering on a green board.

Correct Answer is d

(Signs providing information about destinations and information are green with white letters or symbols)

4. **Only when _____, should you consider driving below the posted speed limit:**

    A. Driving conditions are worsening.

    B. The drivers around you are driving slower.

    C. Driving conditions are getting better.

    D. You are on a road with five-lanes.

Correct Answer is a

(You should only drive under the prescribed speed limit when driving conditions are worsening. This includes the road being slippery or wet, low visibility, or any conditions that hinders your ability to drive the speed limit)

5. If other drivers around you aren't expecting you to stop, turn or slow down, you should first:

A. Press on your brakes several times quickly.

B. Engage the vehicle's emergency brakes.

C. Turn your head to look over your shoulder to locate your blind spot.

D. Honk and let the others know that you'd be stopping, turning or slowing your vehicle.

Correct Answer is a

(When slowing down, turning or stopping, alert other drivers by lightly pressing on your brakes several times)

6. In the case where traffic around you prevents you from crossing through a set of railroad tracks, you should consider proceeding only when:

   A. The train is moving slow and will not be a danger.
   B. The other side of the track is empty and there is enough room.
   C. More than half of your vehicle has already crossed.
   D. There are no trains nearby.

Correct Answer is b

(You shouldn't consider crossing railway tracks in any circumstances. The only exception here is when you are sure that your entire vehicle will clear through all of the tracks and there is enough space to do so)

7. Before passing other vehicles you must:

   A. Alert other drivers by flashing your headlights.
   B. Warn the driver ahead by turning on your flashers.
   C. Show other drivers that you are changing lanes by initiating the proper turn signal.
   D. Blow your horn for 3 seconds to get attention.

Correct Answer is c

(Whenever you are considering passing a driver ahead of you, you should first ensure that the passing lane is clear. Then, you should initiate the proper turn signal and let other drivers know you'll be changing lanes)

8. If you ever get involved in a traffic collision, you are obligated to submit a complete written report in the case:

A. Either you or the other drivers are injured.

B. The collision leads to property damage over a thousand dollars or if there are any other significant injuries.

C. You are the only one at fault.

D. You are obligated to submit a complete written report regardless of the casualty.

Correct Answer is b

(Every driver involved is required to submit a complete written report for the incident if and when a traffic collision leads to more than a thousand dollars in damage, minor or major injury, or an instance of death)

9. **You should consider __, in case you are traveling under forty miles an hour on the highway.**

   A. Driving near the shoulder of the lane.
   B. Turning on your high beam lights.
   C. Blowing your horn to let others know.
   D. Turning on your four-way flashing lights.

   Correct Answer is d

   (In case you have to drive below forty miles an hour on the highway, you should turn on your four-way flashing lights to warn the other drivers behind you)

10. **When crossing through heavy traffic, the minimum amount of space needed to do so largely depend on:**

    A. The weather and road conditions along with the oncoming traffic conditions.
    B. The stop sign near the traffic.
    C. Your turn signals while driving through.
    D. The amount of traffic behind you.

Correct Answer is a

(Spatial requirements for when you have to cross through traffic largely depend on the road and weather conditions, and also the nature of oncoming traffic)

11. If you're driving in your lane and an oncoming vehicle is heading towards you, you should consider:

   A. Blowing your horn, steering to your right, and increase speed.
   B. Sounding the horn, steering to your left, and stopping.
   C. Blowing your horn, steering to your right, and begin applying your brakes.
   D. Sounding your horn, staying within your lane, and braking to alarm the oncoming vehicle.

Correct Answer is c

(If another vehicle ever approaches you head-on in your lane, you should consider sounding your horn to get their attention. If the other driver doesn't budge, you should attempt to escape to your right by steering and braking)

12. **If you're faced with a situation where there's an oncoming car to your right and a bicyclist to your left, you should consider:**

    A. Pulling onto the road's shoulder.
    B. Splitting the difference between them both.
    C. First let the car pass by, and then pass by the bike.
    D. Pass by the bicycle as quickly as you can, then let the car pass by.

    Correct Answer is c

    (In case there's a bicyclist on your left and your intentions are to pass by an oncoming car to your right, don't try and squeeze between them both. Instead, you should let the oncoming car pass by, and then pass by the bicyclist afterwards)

13. **You are driving 55 mph on a freeway and the traffic is moving at 60 mph. You are legally allowed to drive:**

    A. At 60 mph or faster to keep up with the traffic.
    B. Slightly faster than the posted speed limit.
    C. Never faster than the road assigned speed of 60 mph.
    D. Always under the road assigned speed limit between 50 mph and 55 mph.

Correct Answer is c

(You must drive no faster than the posted speed limit)

14. When you are driving in traffic, it is safer to drive by:

    A. Keeping alert and increasing your speed.
    B. Considering the flow of traffic and driving a little faster.
    C. Considering the flow of traffic and driving a little slower.
    D. Driving within the natural flow of traffic.

Correct Answer is d

(You should always drive with the traffic flow and monitor your speed)

15. One of these vehicles must always stop before attempting to cross any railroad tracks:

    A. Trucks and tankers marked with hazardous material placards.
    B. Pickup trucks towing cars, boat trailers or mobile homes.
    C. Vehicles weighing over five thousand pounds.
    D. All the above vehicles.

Correct Answer is a

(Trucks carrying hazardous loads must always stop before attempting to cross any railroad tracks)

16. **Unless mentioned specifically, every residential area has an upper speed limit of:**

    A. 40 mph.
    B. 25 mph.
    C. 35 mph.
    D. 20 mph.

Correct Answer is b

(The speed limit in residential and business areas are 25 mph unless a different speed limit is posted)

17. **If you obtain a Class C Driver's license. You are allowed to drive:**

    A. A three axle vehicle with a Gross Vehicle Weight less than six thousand pounds.
    B. Any three axle vehicle irrespective of its weight.
    C. Vehicles pulling more than two trailers.
    B. An eighteen wheel vehicle pulling three trailers.

Correct Answer is a

(After obtaining a Class C driver's license, you are allowed to drive a three-axle vehicle weighing under six thousand pounds)

18. If a traffic light is found to not be working, you must always:

A. Stop, look to see if other cars are stopped, then proceed if it's safe to do so.
B. Stop as you enter the intersection then continue.
C. Slow down or stop if you don't see other vehicles.
D. Consider the traffic light as a yellow signal.

Correct Answer is a

(If traffic signal is faulty, you should consider it to be a stop sign. You must proceed cautiously)

19. If you happen to be driving on a wet roadway, which of the following should you consider?

    A. Your tires become less effective as you increase speed on the roadway.
    B. Good tires are not affected by the presence of water on a road's surface.
    C. Roads with deeper waters are less dangerous compared to shallow waters.
    D. The roadway tends to become more slippery as you decrease your speed.

Correct Answer is a

(Your tires tend to be rendered less effective as you drive faster on a roadway. When going fast, tires aren't as efficient in wiping down water from a road and lose their grip. In such a scenario you could also begin to hydroplane when the tires lose their grip entirely)

20. You just sold your vehicle. You must notify the New Jersey MVC within ____ days.

    A. 10
    B. 25
    C. 15
    D. 20

Correct Answer is a

(When you sell or transfer a vehicle, you must notify the MVC within 10 days)

21. Which of the following cases should honk your honk?

    A. When you are driving by a crossroad.

    B. When you are passing a motorcycle.

    C. When you spot a young child running into the street.

    D. When you are considering parallel parking.

Correct Answer is c

(Your horn must be treated as a means to warn others where you think they are not seeing you and could potentially be in danger)

22. As you drive towards an intersection, the traffic light turns from green to yellow. In such a case, it's best if you:

    A. Speed up to beat the red light.

    B. Apply the brakes sharply to stop.

    C. Be prepared to stop in the center of the intersection.

    D. Be prepared to stop before the intersection.

Correct Answer is d

(The yellow light is meant to indicate that the light will soon transition to red. In such a case you must be prepared to stop for the red light)

23. If you see someone with a sign or flag at a road construction site ahead, you must:

A. Pay attention to the yellow cones on the road ahead of you.
B. Their actions conflict with what the existing road signs, signals, or laws say.
C. Follow their instructions.
D. In case you are driving under thirty five miles an hour.

Correct Answer is c

(Flaggers are there to keep you safe and drive carefully through work zone areas)

24. You must obey which one the most?

A. A consistent red light.
B. A police officer.
C. A red arrow.
D. A flashing amber.

Correct Answer is b

(The instructions that active traffic police personnel gives outweighs existing signals and signs)

25. If you are driving and see a road sign that is diamond-shaped, what is it implying?

A. There is a road hazard ahead.
B. The sign signifies that you are on an intersection route.
C. There is a bus stop ahead.
D. This is the maximum speed for the road.

Correct Answer is a

(Drivers are warned of special conditions or hazards on the road ahead through diamond-shaped signs on the road)

26. Which of the following is the most effective means to reduce any case of injury or worse getting killed in a traffic accident?

A. Wear your seat belt properly.
B. Only drive on weekdays as necessary.
C. Always staying on the left side of the street.
D. Only driving between two 'o clock and five 'o clock in the evening.

Correct Answer is a

(The best and effective way to reduce the risk of injury or death during a car crash is to wear your seatbelt)

27. What can be the outcome of tailgating a driver in front of you? This is when you drive very close to the other vehicle.

    A. You can risk getting into a road rage incident.

    B. Tailgating can lead to a traffic ticket.

    C. You can help increase the flow of traffic.

    D. All of the Above.

Correct Answer is a

(Tailgating is the one cause behind most rear-end collisions. To avoid such an instance, you can follow a 'three second rule.' Here, you should check as the vehicle passes by a certain point, then count to three and see if you pass the same point. If you do, then you are probably following too closely. Here, it is suggested to slow down so as to not tailgate the other vehicle)

Understanding the Rules of the Road

28. Which area on the road can be labeled as your blind spot?

   A. A spot that you cannot look at without needing to turn your around.
   B. A spot that is behind the vehicle.
   C. A spot that you can see on the left side of your vehicle
   D. A spot that you can see in your right side mirror.

Correct Answer is a

(While driving, there are spots over your shoulder that you can't see without turning your head. These are called blind spots)

29. When backing up, you should first:

   A. Check your mirrors then proceed.
   B. Turn your lights on and off to alert others.
   C. Lower your window then proceed.
   D. Look through the rear window by turning your head around.

Correct Answer is d

(You should not solely depend on your mirrors, window or lights when backing up your vehicle)

30. When you are driving on wet roads and highways, you should:

A. Drive the posted speed limit.

B. Drive above the posted speed limit.

C. Drive a few miles below the speed limit.

D. Tailgate the vehicle ahead.

Correct Answer is c

(Tires tend to lose traction on slippery roads. It is best that you reduce your speed under the prescribed limit to be safe)

31. While all of the enlisted practices are dangerous to carry out while driving, some are actually illegal. Which one is it?

A. Listening to music via earphones or headphones that cover both the ears.

B. Fixing your outside mirrors while driving.

C. Travelling with an unrestrained animal while driving the vehicle.

D. All of the abovementioned options.

Correct Answer is a

(It's likely for people with good hearing to not be able to hear well when a radio is loud. However, it is unlawful to drive while wearing headphones)

32. If you drive on slippery road, you should:

    A. Turn your vehicle slower than normal.

    B. Be quick with changing lanes.

    C. Increase your speed.

    D. None of the above.

Correct Answer is a

(When roads are slick and slippery, you should be especially cautious with how you turn and approach as compared to regular instances)

33. If you're at a railway intersection and the approaching train is close enough or seems to be going fast, you should ideally:

    A. Slow down so you don't damage your tires.

    B. Only attempt to cross the tracks when the train passes completely.

    C. Take the risk and cross the road.

    D. Find another way to cross the tracks.

Correct Answer is b

(Only when you are certain of your safety should you consider driving through a railway intersection)

34. You are traveling down a one-way street. You may turn right onto another one-way street only if:

   A. There's a sign permitting you to make the turn.

   B. Cars are moving to the left.

   C. Traffic on the street is moving to the right.

   D. Cars on the right are stopped.

Correct Answer is c

(Only when there is no sign prohibiting you to turn into a one-way street to your right can you turn into a one-way street on your right)

35. When you see a vehicle approaching you with its high beams turned on, you should consider looking towards __ side of the road:

   A. Both.

   B. The centerline.

   C. The right.

   D. The rear.

Correct Answer is c

(When a vehicle approaches you with its high beams on and refuses to dim them, you should try looking towards the right side of the road)

36. When you're driving down a road and it's marked with a solid yellow line and a broken yellow line to your side, you're only allowed to pass:

    A. Only if it's urgent.

    B. If you are driving on the highway.

    C. If there's no traffic in that direction.

    D. If you are at a junction.

Correct Answer is c

(If the road's center has a broken yellow line and a solid yellow line, with the broken line next to the lane you're driving in, you may cross the lines to pass only when there's no upcoming traffic)

37. What does a rectangular-shaped sign mean?

  A. It is a sign denoting a crosswalk.

  B. The sign means that there's a railroad ahead.

  C. It means that it's a yield sign.

  D. The sign indicates a speed limit.

Correct Answer is d

(Signs that inform drivers of road regulations including speed limits are generally rectangular in shape)

38. You are driving and approached a railway. You don't see any warning lights or a train in either direction. What is the speed limit that you should drive at in this case?

  A. 15 mph

  B. 40 mph

  C. 20 mph

  D. 25 mph

Correct Answer is a

(When approaching a railroad crossing and unable to spot any train on either side, you are allowed to drive through under 15 miles per hour)

39. You were driving along a road and ended up getting involved in a minor collision with a parked vehicle with no owner around. What should you do?

A. You should act responsible and leave your information on their vehicle.
B. Call 911 and report the accident.
C. You should do both A and B.
D. In case the damage is under a thousand dollars, you are not obligated to report it.

Correct Answer is c

(If you ever get in such a situation, you are obligated to leave a note with your name, phone number, and address attached to the vehicle you accidentally damaged. Further, you must also report the incident to the local city police)

Chapter 3
Recognizing Road Signs

Learning how to drive can be like learning a new language, especially when road signs are involved. Road Signs are necessary for directing traffic, warning drivers of upcoming road conditions, and keeping pedestrians and driver's alike safe. Chapter 3 will present each type of road sign, and it is critical you memorize the meaning of each one of them. Road signs are distinguishable by their color, their shape, or a combination of the two. While you may find it tricky to differentiate between two or more of them in the beginning, it will get easier with practice and as you gain more driving experience. Note that you will need to carefully divide your attention between your own vehicle, those around you, as well as the road signs along your route.

The 10 questions in this section will assist you as you familiarize yourself with each type of road signs and various other signals, colors, and shapes you may see while driving. It's best to go over this section multiple times until you feel confident in your abilities to recognize each of these characteristics. Even the road has its own system of communication, and it's now your responsibility to make sure you understand the language.

Let's begin:

1. When you're driving on the center lane and encounter this sign, what are you allowed to do?

    A. You are only allowed to turn left.

    B. You are only allowed to turn right.

    C. You are only allowed to drive the center lane.

    D. All of the above.

Correct Answer is a

(Center Turn Lane: This sign indicates left-turning vehicles from both directions)

2. What does this sign implies?

    A. There's a right turn ahead.

    B. There's an Intersection ahead.

    C. Up ahead is a lane change.

    D. The road merges ahead.

Correct Answer is b

(The shown sign is called a T Intersection sign and alerts you that the road you're on ends ahead. It implies that you should slow down and prepare to halt before turning in either direction)

3. This sign on the right implies:

A. There is a divided road ahead.
B. There are vehicles passing ahead.
C. There's an additional traffic lane ahead.
D. There is lane crossing ahead.

Correct Answer is c

(The sign is called an Added Lane sign. It alerts a driver that an additional driving lane is going to be added to the main roadway for traffic entering from a side lane)

4. When you see the sign, what must your immediate concern be?

A. You must be concerned with the type of vehicle you have.

B. You could damage your tire.

C. Law enforcement issuing you a citation.

D. You could lose control of your vehicle if you end up drifting off to the shoulder since its lower.

Correct Answer is d

(Shoulder Drop Off: This sign warns that the shoulder is lower than the road. If you drift off of the road, don't panic, tightly hold the steering wheel and turn back on the road at a lower speed)

5. You're driving through a road and see the sign on the right. What does it mean?

A. It implies that you shouldn't drive.

B. It implies that the road is slippery when wet.

C. It means that the road curves ahead.

D. It alerts you that the road it approaching a hill.

Correct Answer is b

(The sign alerts drivers that the road is slippery when it gets wet. Proceed with caution)

6. The sign on the right means:

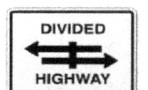

A. It is compulsory for you to turn either right or left.
B. You are approaching a fork in the road.
C. The road you're on is soon to approach a divided highway.
D. The sign indicates a divided bridge.

Correct Answer is c

(The sign here is called a Divided Highway sign. It implies that the current road shall soon intersect with a divided highway which consists of two one-way roads separated by a guide rail or a median)

7. What do rectangular-shaped signs signify?

A. It signifies that there's a school crossing ahead.
B. It signifies that there is road construction ahead.
C. It means that you should slow down.
D. They are speed limit signs.

Correct Answer is d

(Rectangular signs generally convey the speed limit. These are mostly colored white and have black letterings)

8. What does a diamond-shaped sign imply?

A. The possibility of a road hazard up ahead.
B. The road is part of a one way street.
C. You are approaching a crosswalk.
D. This is a secondary speed limit sign.

Correct Answer is a

(A diamond-shaped sign is a road hazard sign and colored yellow. These invoke caution and convey passersby of a hazard ahead)

9. What does this sign mean?

A. U-turns are prohibited.
B. You may turn left only if it's safe.
C. Right turns are not allowed at the intersection.
D. No turns are allowed at this intersection.

Correct Answer is c

(If you see this sign at an intersection, right turns are prohibited)

10. You encounter the sign to the right while driving through a road. What does it imply?

   A. There's a sharp turn to the right coming up ahead.
   B. It's alerting you of a double curve, one that first turns right, then left.
   C. It's alerting you of a double curve, one that first turns left, then right.
   D. It signifies that the pavement is going to end ahead.

Correct Answer is b

(The sign shown is a double curve. It stands to warn the driver that the road will first curve to the right and then to the left. It is to let drivers know that they should approach it at a safe speed)

Chapter 4
Traffic Lights & Signals Awareness

Chapter 4 will cover Traffic Lights and Signals Awareness, which are necessary for keeping traffic moving in an orderly, uniform fashion. All drivers must obey these signals at all times unless an emergency necessitates another action, or a law enforcement officer directs traffic to do otherwise. Failure to follow traffic lights correctly could result in an accident at worst, and confusion for other drivers at a minimum. Small differences, such as whether the light is solid or flashing, must be observed. It's vital that you are just as familiar with commonplace traffic lights as you are with more unique signals like those that flash red on a school bus stop sign.

Keep in mind traffic lights will not protect you from making mistakes, nor will they keep you safe when other drivers fail to adhere to them. You will still need to practice defensive driving and prepare for situations in which the traffic lights are inoperable. Although the light may tell you a certain action is safe to take, you'll still need to look around and check your surroundings for other cars, hazards, and pedestrians before proceeding. There are 13 multiple-choice questions in this section through which you can test your knowledge of traffic lights and signals.

Let's begin:

1. Is it legal to proceed immediately if you are stopped at an intersection and the traffic light has just turned green?

    A. Yes, you can immediately proceed.
    B. Yes, but yield to any remaining vehicles or pedestrians who are still in the intersection.
    C. Yes and No, other vehicles and pedestrians in the intersection must yield to you.
    D. No, you should wait thirty seconds.

Correct Answer is b

(Yes, but yield to any remaining vehicles or pedestrians who are still in the intersection)

2. In which of the following situations is it illegal to enter an intersection?

A. You cannot cross and intersection while causing traffic congestion on both sides.
B. The lane you wish to use is closed.
C. The traffic light just turned yellow.
D. Cars on the right are entering the intersection.

Correct Answer is a

(You cannot enter an intersection if it results in obstructing traffic. If you block the intersection, you can be cited)

3. What does a flashing yellow light mean?

A. Train is coming.
B. Slow down and proceed with caution.
C. Work zone.
D. Speed up to beat the red light.

Correct Answer is b

(A flashing yellow traffic light indicates that you should drive with caution)

New Jersey Driver's Practice Tests

4. In which of the following situations should you stop?

    A. When the red light is flashing.

    B. When the traffic light is a steady yellow.

    C. When the traffic light shows a yellow arrow.

    D. When the yellow light is flashing.

Correct Answer is a

(A flashing red traffic signal has the same meaning as a stop sign. You must come to a complete stop, look both ways then proceed only if it's safe to do so)

5. When a school bus in front of you stops and displays red lights, what does it mean?

    A. If there are no children on the lane, you can proceed.

    B. You are not permitted to proceed when the red lights are blinking.

    C. If you are facing the back of the bus, you can proceed.

    D. If it's on the same side of a divided highway, you are permitted to proceed.

Correct Answer is b

(When a stopped school bus is blinking its red lights, you may not pass until the red lights stop flashing)

6. Which of the following situations are you allowed to drive with just your parking lights on?

    A. 20 minutes before sunset or 20 minutes after sunrise.
    B. Never.
    C. On days where there's less traffic.
    D. During rainy days with poor visibility.

Correct Answer is b

(Under no circumstances can you drive using only your parking lights. It would be difficult for other vehicles to see you at night and can result in a collision)

7. You are entering a train crossing that isn't equipped with signals, what is the most appropriate action that you should take?

    A. You should turn around since it's dangerous.
    B. Drive slowly and expect to come to a stop.
    C. Keep driving, there's no need to stop.
    D. If you don't see any lights or gates speed up.

Correct Answer is b

(Slow down and expect to wait if you come upon an unmarked train crossing. Make sure that there are no trains coming from any direction. If you see or hear a train approaching, come to a complete stop and wait until the train has passed)

8. You may turn right on red if you:

    A. Stop first and check for traffic and pedestrians.
    B. If you have a red arrow pointing right.
    C. If you are driving in the left lane.
    D. If you slow down your vehicle first.

Correct Answer is a

(You may only turn if it is safe to do so and if there is no sign prohibiting turns on a red light. Be careful of pedestrians crossing in front of your vehicle)

9. Which actions should you take when a yellow arrow emerges as you are about to make a left turn from a designated left turn lane?

   A. Do not take the turn under any situation.
   B. Speed up to beat the traffic.
   C. Wait and be ready to follow the next traffic signal that appears.
   D. You have the right-of-way over other vehicles.

   Correct Answer is c

   (Before heading in the direction indicated by the sign, plan to brake and yield the right-of-way to oncoming traffic)

10. Which of the following conditions should be met before you can proceed cautiously through a yellow light?

    A. If an emergency vehicle is crossing your lane.
    B. If pedestrians are not crossing.
    C. If you are about to make a right turn.
    D. If you're already in the middle of an intersection.

Correct Answer is d

(If you're already in the middle of an intersection and a yellow light emerges, you must proceed cautiously into the intersection)

11. Your red light has turned green, but there are still other vehicles on the intersection. What should you do?

   A. Since the light is green, you may enter.
   B. Only proceed if you can easily get through the other vehicles in a safe manner.
   C. Before entering, wait until all of the vehicles have passed through the intersection.
   D. Since you have the right-of-way, proceed across the intersection.

Correct Answer is c

(Before proceeding, you must wait until the intersection is clear)

Traffic Lights & Signals Awareness

12. When a traffic signal is broken or not working, which of the following actions can you take?

    A. Slow down and wait for it to be fixed.

    B. Find a new route.

    C. Proceed as if it were a four-way stop sign.

    D. Go about your business as usual.

Correct Answer is c

(An intersection with a broken traffic signal should be treated as a four-way stop)

13. You are at an intersection of two-way streets and you have come to a full stop at a red traffic light. What should you do now?

    A. If the road is clear, drive straight ahead.

    B. If the road is clear, turn left.

    C. If the path is open and clear, turn right, unless otherwise indicated.

    D. You can do any of the above.

Correct Answer is c

(Turn right if the way is clear, unless otherwise posted)

Chapter 5
Turns & Intersection Maneuvers

Being a safe driver means staying in your lane, communicating your intentions with other drivers, and paying attention to what others are doing on the road. In Chapter 5, we'll be discussing making turns and how to maneuver your vehicle lawfully and safely at an intersection. Making a turn doesn't have to be complicated, and the most important part is making sure you're letting other drivers know what you're planning to do. While your vehicle's lights and signals are ideal when you need to tell other drivers which direction you intend to move in, know that some drivers will use hand signals instead. It's important to be aware of these different methods of communication as being unfamiliar with them can quickly create a dangerous situation on the road. Get used to using your vehicle's lights and signals while driving, even if it's a simple lane change.

The 19 questions in this section will test your knowledge about having the right-of-way, yielding to other vehicles, reacting to emergency vehicles, and responding at an intersection. Remember that you will be faced with new situations daily while driving. Always choose the safest option and communicate your intentions with other drivers.

Let's begin:

1. In which of the following situations should you yield the right of way to an approaching vehicle?

   A. When you are already pass traffic on the left.

   B. When you are already in an intersection.

   C. When you are proceeding in the straight path.

   D. When you are taking a left turn.

Correct Answer is d

(When two vehicles enter an intersection at the same time from different directions, the driver turning left must yield to traffic. If the vehicle driving straight enters the intersection where another vehicle is already turning left, they must wait for that vehicle to finish its turn before proceeding through the intersection)

New Jersey Driver's Practice Tests

2. You are at an intersection; the light is green and you wish to take a left turn but there is heavy oncoming traffic. Which of the following actions should you take?

   A. Use the next U-turn.
   B. If traffic doesn't clear, get in your right lane.
   C. Be patient and wait in the center of an intersection for the traffic to clear.
   D. Proceed since you have the green light.

Correct Answer is c

(When taking a left turn in front of oncoming traffic, you must wait until oncoming traffic to passes before turning. If the light is green and no other vehicle ahead of you decides to make a left turn, you can enter the intersection to prepare for your left turn)

3. A truck is turning right with two open lanes in both directions, what needs to be done?

   A. The truck may complete their turn in the left or right lane.
   B. To complete the turn, the trucks often have to use a portion of the left lane.
   C. The trucks need to be in the right lane always.
   D. The trucks may do all of the above.

Correct Answer is b

(Trucks tend to move slightly to the left lane when making a right turn. This maneuver gives the driver more room on the right side of the road. Be cautions when driving next to trucks)

4. When you need to take a right turn at an intersection, you should decrease your speed and:

- A. Get prepared to shift to the left lane early.
- B. Should not drive in the bicycle lane.
- C. Use your turn signal at least 100 feet before turning.
- D. Turn from the middle lane where it's safer.

Correct Answer is c

(You should start signaling about 100 feet before the turn when making a right turn)

5. You wish to make a right turn. Which of the following positions should your car be in?

   A. Close to the middle of the driveway.

   B. Near the street's left side.

   C. Near the street's right side.

   D. Beyond the middle of the junction as you begin to turn.

Correct Answer is c

(When making a right turn, get as far to the right side of the road as possible while staying in your lane)

6. From a private road, you wish to enter a roadway. What should you do?

   A. Stick the nose of your car out in traffic.

   B. Come to a complete stop with a portion of the vehicle on the roadway to alert all other drivers.

   C. Drive into the roadway quickly so you can blend seamlessly with the traffic.

   D. Yield the right-of-way to pedestrians and vehicles on the roadway when approaching a roadway from a driveway or private lane.

Correct Answer is d

(You must yield the right-of-way to pedestrians and vehicles on the roadway when approaching a roadway from a driveway or private lane)

7. **The left arm and hand of the driver are extended downward. This hand gesture indicates that the driver intends to do the following:**

    A. Get the engine going.
    B. Take a right turn.
    C. Come to a halt.
    D. Take a left turn.

Correct Answer is c

(As a driver extends his or her left arm and hand downward, they are signaling that they want to come to a halt. If you're behind a driver who uses hand signs, keep a safe distance)

New Jersey Driver's Practice Tests

8. You wish to enter traffic from a stop away from the curb. Which of the following conditions should you follow?

A. You should drive slower than the other vehicles.

B. You'll need a big interval to catch up to the traffic's speed.

C. You should only pull into the lane after the first three vehicles have passed.

D. You should switch on the hazard lights and approach the traffic cautiously.

Correct Answer is b

(You need a wide enough distance to get up to speed, access the flow of traffic, and not obstruct it when entering the traffic from a stopped position)

9. Which of the following actions should you take while waiting in the intersection to complete a left turn?

A. Put on your hazard lights to alert others.

B. Make sure to keep your wheels straight while signaling appropriately.

C. Press your brakes a few times to alert others.

D. If a car is in the way, drive around its rear.

Turns & Intersection Maneuvers

Correct Answer is b

(When waiting to make a left turn, keep your wheels straight. This could protect you if another driver approaches you from behind)

10. While preparing to make a right turn, which of the following actions should you take?

  A. No need to signal since you have the right of way.
  B. Before entering the right lane, come to a complete stop and prioritize other traffic.
  C. If required, slow down or stop, and then turn to the right.
  D. Switch on your right signal, slow down, then stop.

Correct Answer is c

(When making a right turn, signal first, then slow down and complete the turn)

11. Which actions should you take when a car is attempting to merge into your lane while you are driving on an interstate (freeway)?

   A. Try to provide space for the merging vehicle safely, if possible.
   B. Get out of the way of merging traffic by quickly speeding up.
   C. Stop quickly to allow vehicles to merge.
   D. Do none of the above, just drive as usual.

Correct Answer is a

(If at all practicable, leave space for cars approaching a highway. Merge into the next lane to make room for oncoming traffic. If you are unable to merge, slow down to allow cars to join traffic as quickly and safely as possible)

12. To turn right from a highway into two lanes in your direction, turn from:

   A. The lane nearest to the road's middle.
   B. The lane closest to the road's edge or curb.
   C. According to the oncoming traffic, any direction.
   D. None of the above options.

Correct Answer is b

(Start and finish your turn in the lane closest to the right hand curb. Do not swing deep into the other lane)

13. When should you initiate your turn signal at the next intersection, if you wish want to turn right?

   A. A minimum of 30 feet prior to the turn.
   B. Just before you get to the intersection.
   C. A minimum of 100 feet prior to the turn.
   D. When you see cars approaching from ahead.

Correct Answer is c

(You should start signaling about 100 feet before the turn while making a right turn)

14. Which of the following actions should you take when you reach an intersection that is congested do to other traffic?

   A. Slow down until the traffic passes.
   B. Make a U-turn from the middle lane.
   C. Before entering wait until you can drive through the intersection completely.
   D. Both B and C.

Correct Answer is c

(If traffic is congested on the other side of the intersection you won't be able to pass through. If you can't get across the intersection entirely, wait until traffic clears so you don't obstruct the road)

15. In which of the following situations do you have the right-of-way?

A. When you are about to enter a roundabout.

B. When you are entering a parking garage.

C. When you are on the left side of the road.

D. When you've already entered an intersection.

Correct Answer is d

(Before entering an intersection drivers must yield the right-of-way to those already in the intersection)

16. **Which of the following actions should you take when you cannot immediately see the emergency vehicle after hearing a siren while driving?**

    A. Do nothing since you can't see the vehicle.
    B. Check to see if it's on your street by pulling up to the curb.
    C. Decrease speed and wait.
    D. Follow the path of the emergency vehicle.

    Correct Answer is b

    (If you hear a siren but are unsure the location of the emergency vehicle, pull off to the right side of the road and wait until you are certain it is not approaching you)

17. **Which of the following actions should you take when traffic is blocking the intersection even though you have a green light?**

    A. Wait until the traffic clears and stay out of the intersection.
    B. Since the light is green, you may proceed.
    C. Change lanes to try to get past the traffic jam.
    D. Make use of the exit lane.

Correct Answer is a

(Wait until the traffic clears and stay out of the intersection. You should not enter an intersection until you are able to pass through it)

18. You see a red traffic light blinking at an intersection. What does it mean?

    A. Before entering, decrease your speed.

    B. Before entering, come to a complete stop.

    C. Come to a complete stop and wait for the yellow light.

    D. None of the above.

Correct Answer is b

(If a blinking red light appears at an intersection, drivers must come to a full stop, yield to all vehicles and pedestrians, and proceed only when it is safe to do so)

19. Two vehicles are about to enter an intersection at the same time from different highways. The vehicle that must yield the right-of-way is:

   A. Any of the vehicles have the right-of-way.

   B. The vehicle to the left has the right-of-way.

   C. The vehicle on the right has the right-of-way.

   D. Both vehicles have the right-of-way.

Correct Answer is b

(When two vehicles arrive at an intersection at the same time and are monitored by signs or signals, the driver on the left must yield the right-of-way to the driver on the right)

Chapter 6
True or False Trivia

There are a vast number of signs you'll see on the road while driving, and it can be hard to learn the difference between each of them. This is because some road signs are very similar looking, whether because of their color, shape, or some other factor. Chapter 6 is a review of many of these road signs, just to make sure you have them memorized. Once you're on the road, you won't have time to look up what a sign means as you pass by it. You'll need to already know and then respond accordingly. It's better to learn them now so you'll recall their meanings easily later on.

The ability of traffic to flow safely and efficiently depends greatly on each driver's knowledge of every road sign and it also prevents accidents from occurring. Even if a road sign doesn't apply to you, get used to noticing them while driving and recalling their meanings as you see them. Doing so will help ensure you don't forget the road signs over time, as you'll eventually need to know what all of them mean

True or False Trivia

in different driving situations. There are 10 True-or-False Trivia questions in this section that will assess your ability to recognize these various types of road signs.

Let's begin:

1. The following road sign can be identified as a 'Curves on the Road' sign.

   A. True
   B. False

Correct Answer is a

(This road sign can be identified as a 'Curves on the Road' sign. The path ahead of you has a series of curves. The path will bend to the right first, then to the left)

2. The following road sign can be identified as a 'Steep Hill Ahead' sign.

   A. True
   B. False

Correct Answer is a

(This traffic sign alerts you about an upcoming hill on the route. To control speed and conserve brakes, slow down and be prepared to change to a lower gear)

3. This sign indicates that another road will be entering your road.

   A. True
   B. False

Correct Answer is a

(This is a Road Entering Curve sign. A side lane enters from the right as the main road bends to the left. Take extra caution while approaching the intersection)

4. The following road sign can be identified as a sign which indicates that only Fire Trucks are allowed to make a U-turn from this lane.

   A. True
   B. False

Correct Answer is b

(This is a U-turn only sign. This signal requires all traffic to make a U-turn, not just Fire Trucks)

5. **This road sign can be identified as a 'T-Intersection ahead' road sign.**

    A. True
    B. False

Correct Answer is b

(This is a road sign used to indicate merging traffic. It's possible that traffic from another road is merging into your path. Enable other traffic to cross into your lane or change lanes if necessary)

6. **The following road sign can be identified as a sign which warns the drivers that they must stay on the pavement.**

    A. True
    B. False

Correct Answer is a

(This is a soft shoulder sign. It is used to warn drivers to stay on the pavement due to the shoulder being made up of dirt, rocks or a mixture of both)

7. There are broken white lines on the pavement as the accompanying picture shows. It means that passing is prohibited.

   A. True

   B. False

Correct Answer is b

(If you see a broken white line, that means you can pass or change lanes according to the safety of the situation)

8. The following road sign can be identified as a 'Roundabout Ahead' sign.

   A. True

   B. False

True or False Trivia

Correct Answer is a

(This road sign represents a 'roundabout ahead' sign. A roundabout is a circular intersection without a traffic light where traffic travels in a counter-clockwise directional circle around a central area)

9. **The following road sign can be identified as a 'School Zone' sign.**

    A. True
    B. False

Correct Answer is b

(This road sign is a Pedestrian Crossing sign. Keep an eye out for pedestrians entering the roadway. If possible/necessary, slow down or stop)

10. **This is a 'Two Lane Only' road sign.**

    A. True
    B. False

Correct Answer is b

(This road sign is identified as a Two-Way Left Turn sign. A two-way left-turn lane is a lane in the middle of the road that is designated for cars turning left from the road in both directions)

Chapter 7
Driving Under the Influence

There are many factors that can influence the decisions you make while driving. These include your mental state, alcohol consumption, drug use, and even the amount of sleep you get, among others. In Chapter 7, you'll learn about the risks involved when driving under the influence of drugs or alcohol. We will be discussing the myths that surround reducing your blood alcohol content (BAC) and taking a more in-depth look at how alcohol affects both your mind and body negatively while driving. You'll even get to learn how seemingly harmless substances like cold medicine and non-prescription drugs can affect your ability to operate a vehicle safely.

Finally, you'll learn about the consequences of driving under the influence and how you can best avoid them. Not only can driving under the influence result in losing your driver's license, but it can also cause an accident that may prove to be fatal for yourself or others. The good news is that much of this is preventable with some simple steps to keep yourself and others safe. The 15 questions in this section will prepare you with the knowledge you need to operate your vehicle with the best state of mind possible.

Let's begin:

1. Which actions should be taken when a driver has taken a non-prescription drug?

    A. Prior to driving read the labels.

    B. Drive only during night hours.

    C. Drive only on the weekends.

    D. Non-prescription drugs don't have any effects.

Correct Answer is a

(Non-prescription drugs have the potential to impair your ability to drive safely. Before starting a car, first read the label for any warnings about the consequences)

2. If alcohol and another drug are mixed in your blood, what effect do you think they'll have?

    A. Boost all their effects.

    B. Have little bearing on the ability to drive.

    C. Reduce the drug's or medicine's side effects.

    D. Diminish the alcohol's effects.

Correct Answer is a

(A combination of alcohol and other drugs severely reduces your ability to drive and can cause serious health problems, which can include death)

3. Which of the following is a factor that blood alcohol content DOES NOT depend on?

   A. The amount of time between drinks.
   B. The amount of alcohol you consume.
   C. The weight of your body.
   D. Your level of workouts in the gym.

Correct Answer is d

(Your blood alcohol concentration is determined by the amount of alcohol consumed, the amount of time between drinks, and your weight)

4. What is the impact of consuming alcohol while either taking a prescription or over-the-counter medication?

    A. The drug will aid in the treatment of sore throat.

    B. The drug counteracts the alcohol's effects.

    C. The effects of the alcohol could be amplified by the mixture.

    D. Both B and C.

Correct Answer is c

(When alcohol and other substances are consumed together, the symptoms of both are amplified)

5. What medications, aside from alcohol, can impair your driving ability?

    A. Over the counter pain medication.

    B. A treatment for a cold.

    C. THC products such as Marijuana.

    D. All of the above options.

Correct Answer is d

(Driving when under the influence of alcohol or other substances combined with alcohol is a criminal act)

6. What can happen if caught driving while drunk or under the influence of other drugs?

A. Revocation of a driver's license.
B. A compulsory fine.
C. The possibility of imprisonment.
D. All of the above options.

Correct Answer is d

(If you drive while under the influence of alcohol or drugs, your license may be suspended, you may be cited, or you may be jailed)

7. Identify a reaction that DOES NOT take place after drinking.

A. Your speech becomes slurred.
B. You have a lower level of alertness.
C. The perception of speed and distance is impaired.
D. You relax so that you can focus.

Correct Answer is d

(Alcohol slows down your response rate, impairs your vision, and alters your perception of speed and distance)

8. An individual under the age of 21 is prohibited from driving with a blood alcohol concentration of:

   A. 5%.
   B. 1%.
   C. 8%.
   D. None of the above options.

   Correct Answer is b

   (When you are under the age of 21, your driving rights could be suspended if you are found guilty of driving with a blood alcohol concentration of 1 percent or more)

9. If you plan to drive after drinking, keep in mind that alcohol impairs your ability to:

   A. Make sound decisions.
   B. See properly and clearly.
   C. Coordinate your actions.
   D. All of the above options.

   Correct Answer is d

   (Alcohol will impair your balance, vision, and judgment by slowing your reflexes and affecting your coordination, vision, and judgment. It also reduces one's ability to adapt quickly to shifting circumstances)

10. The most safe and effective way to lower the blood alcohol content is to:

A. Have a cup of tea.

B. Work out.

C. Give the body enough time to detox from alcohol.

D. Drink lots of water.

Correct Answer is c

(The best way to successfully lower your blood alcohol concentration is to abstain from drinking for a period of time)

11. Which of the following is a factor that contributes to healthy driving if you drink alcohol socially?

A. Move closer to the steering wheel.

B. Have a friend drive instead.

C. Take a warm and long bath prior to driving.

D. Stop drinking 30 minutes before driving.

Correct Answer is b

(Getting a lift home with a friend who does not drink is the safest way to stop driving when distracted or drunk)

12. How does alcohol affect your decisions and driving abilities?

A. It improves driving abilities but impairs judgment.
B. It impairs your driving and decision making.
C. It has little effect.
D. It has little effect on judgment but has a negative impact on driving abilities.

Correct Answer is b

(Many abilities required for safe driving are harmed by alcohol, including decision making, vision clarity, speed and distance judgment)

13. Which selections are affected by alcohol?

A. Sensory motor skills.
B. The amount of time it takes for you to respond.
C. The ability to see clearly.
D. All of these options.

Correct Answer is d

(Alcohol can make you feel less inhibited, but it can also make you more willing to take risks)

14. The effects of alcohol can be determined by:

    A. Your body mass index.

    B. The last time you ate.

    C. The amount of time between drinks.

    D. All three options are correct.

Correct Answer is d

(Blood alcohol content is determined by body weight, alcohol consumption and drinking intervals)

15. If you refuse to take a breath or blood test, what could happen to your driver's license?

    A. You will lose your insurance.

    B. The Police officer will keep your license.

    C. You will lose your driver's license.

    D. None of the above.

Correct Answer is c

(If you refuse to take the test, your license could be suspended)

Chapter 8
Developing Safe Driving Habits

Safe driving is like a muscle you have to develop with time, practice, and experience. Your ability to be a safe driver will be dependent upon the habits you start to build now. In Chapter 8, we'll be discussing some of these habits as well as the actions you should take when presented with various situations while driving. You'll need to learn new skills and then practice them until they're instinctual. In other words, safe driving should become something you're so used to that you don't even need to think about it. You will find yourself in unforeseen circumstances on the road all the time, especially as a new driver, and you need to expect the unexpected.

This chapter will cover good sleeping habits, the effects of alcohol on your system, and how to spot potential problems before they turn into accidents. There won't just be other vehicles like your own on the road. You'll also need to learn how to drive around large trucks, bicycles, slow-moving vehicles, buses, and emergency vehicles. The 20 questions in this section will help prepare you as you begin to develop safe driving habits on the road. Remember that these new skills will get easier to master as you gain more driving experience.

Let's begin:

1. What action must you take when you see flashing lights an emergency vehicle?

    A. Stay in your lane at all times.

    B. Come to a complete stop in the center lane.

    C. Pull over and stop.

    D. Slow down and stay in your lane.

Correct Answer is c

(When you see flashing lights from an emergency vehicle you must pull over, stop then process when it's clear)

2. You see a person approaching the lane ahead, but there is no crosswalk. What should you do?

    A. Tell them to get out of your way.

    B. Blow your horn politely but keep going.

    C. Come to a complete stop and let them pass.

    D. Slow down only.

Correct Answer is c

(Pedestrians in the street must always be given the right-of-way)

3. What happens when the amount of alcohol in your blood increases?

    A. Reduces the speed at which you respond.

    B. It makes you feel insecure.

    C. It begins to decompose itself at a faster rate.

    D. Reduces the chances of making a mistake.

Correct Answer is a

(Alcohol slows your responses, boosts your confidence, and helps you make more mistakes)

4. Which of the following actions should you take if you have a tire blowout?

    A. Apply a heavy braking force to stop the car.

    B. Allow the car to slow down and come to a halt.

    C. Allow free movement of the steering wheel.

    D. Keep going until you get to a tire shop.

Correct Answer is b

(Keep your foot off the accelerator and softly apply the brake if one of the tires blows out. Keep driving straight and look for a safe place to change the tire or call for help)

5. Where should the pedestrians walk on a road which has no sidewalks?

    A. The side of the road with the most traffic.

    B. The side of the road with the least traffic.

    C. Oncoming traffic side of the road.

    D. Outgoing traffic side of the road.

Correct Answer is c

(Pedestrians should walk on the side of the road closest to them, facing the traffic)

6. Ahead of you on your road is a school bus which is flashing lights, what must you do?

    A. Stop only if you see kids leaving the bus.

    B. Reduce your speed to 30 mph.

    C. Come to a complete stop and wait.

    D. Go around the bus and make a U-turn.

Correct Answer is c

(When a stopped school bus flashes its red lights, you must come to a complete stop and wait until the red light stops flashing)

7. Lack of sleep has the same effect on your ability to drive safely as:

   A. Alcohol.

   B. Water.

   C. Food.

   D. Coffee.

Correct Answer is a

(The feeling of being exhausted is the same as being drunk. This will affect the ability to drive safely in a similar way)

8. Motorists should be aware that all bicycles used after dark must have:

   A. Front and rear fenders with yellow reflectors.

   B. Reflectors on the front and back of the bicycle.

   C. Handlebar pegs with traction.

   D. None of the Above.

Correct Answer is b

(A bicycle with a prominent front and rear reflector must be used at night)

9. In which of the following situations should you dim your headlights?

A. If you are behind another vehicle.
B. If someone is behind your vehicle.
C. If your light settings are too high.
D. All of the above options.

Correct Answer is a

(When a vehicle is approaching you or when you are behind another vehicle, you must dim your headlights to a lower beam)

10. Which of the following should a motorist do while approaching a bicyclist?

A. Increase speed to overtake them.
B. Blow your horn to alert them to get out of the way.
C. Move into the opposing lane very quickly.
D. Proceed with caution.

Correct Answer is d

(When passing bicyclists, drivers should slow down because the breeze from your car will quickly disrupt a bicyclist's equilibrium)

11. An emergency vehicle with flashing lights is behind your vehicle while you were driving down a one lane street. What should you do?

    A. Use your emergency hazardous lights.

    B. Safely stop near the side of the road.

    C. Accelerate and take the next available exit.

    D. Decrease your speed until the vehicle passes.

Correct Answer is b

(If an emergency vehicle with flashing lights is behind you on a one-way lane, you must drive to the closest roadside and come to a complete stop)

12. **Which of the following is the most frequent cause of work zone collisions?**

    A. Blow-outs of tires.
    B. Hydroplaning due to sprayed water on the lane.
    C. After driving over wet paint, there was a loss of steering control.
    D. Being Careless and driving too fast.

    Correct Answer is d

    (In work areas, carelessness and driving too fast are the leading sources of traffic-related deaths)

13. **Which of the following is something that you should always do at crosswalks, railroads and intersections?**

    A. Listen and continue as normal.
    B. Keep an eye out on the sides of the vehicle to see what's approaching.
    C. Decrease your speed, look, then proceed.
    D. None of the above.

Correct Answer is b

(When approaching a location where vehicles, trains or pedestrians can cross your route, look to the sides of your vehicle to ensure no one is approaching)

14. In contrast to a motorist, a bicyclist is not expected to:

    A. File a report of all injuries.

    B. Make sure all turns are signaled.

    C. Make sure all turns are signaled.

    D. Obtain insurance for the bicycle.

Correct Answer is d

(Bicyclists must follow the same traffic laws as motorists, but they are not expected to insure their bicycles)

15. What should you do after an emergency vehicle with its siren on passes you?

    A. Get close to the police car.

    B. Increase your speed to see what happened.

    C. Stay at least 500 feet away from the police car.

    D. Call 911.

Correct Answer is c

(After being passed by an emergency vehicle, you must maintain a reasonable distance of 500 feet or more behind the emergency vehicle)

16. What should you do when you hear a fire engine siren?

    A. Reduce the speed until it reaches you.

    B. Use your hazard lights while driving.

    C. Slow down, pull over and stop.

    D. Stop immediately.

Correct Answer is c

(When you hear the siren of an emergency vehicle should slow down, pull off to the edge of the road or to the curb and come to a complete stop)

17. Which of the following factors influences your attention, vision, judgment, and memory?

    A. A blood alcohol content that is over the maximum or legal limit.

    B. No, alcohol isn't one of them.

    C. A small amount of alcohol.

    D. A (BAC) of 0.4%

Correct Answer is c

(Concentration, vision, reasoning, and memory are all harmed by even a small amount of alcohol)

18. If your blood alcohol level is .05, you:

   A. Are 3-7 times more likely to crash than someone who hasn't had any alcohol.

   B. Should drink coffee before driving.

   C. Will be a very safe driver.

   D. You have a blood alcohol limit above the legal limit but are also able to drive.

Correct Answer is a

(It is not safe to drink any amount of alcohol while driving. However, you're three to seven times more likely to get in a car accident than someone who doesn't have any alcohol in their system)

19. Which of the following is the best way to describe what a symbol or an emblem for a slow-moving vehicle would look like?

   A. A sign that is yellow and diamond-shaped.

   B. A sign that is blue and round.

   C. A sign that is a red square.

   D. A sign that is an orange triangle.

Correct Answer is d

(A reflective orange triangle represents a slow-moving car. Vehicles bearing this symbol should be traveling at speed of 25 mph or less)

20. **A huge truck is turning right with two lanes in both directions just ahead of you. Which of the following could be true about the truck?**

    A. When turning, the truck must remain in the left lane.
    B. To complete the right turn, the truck may need to swing wide.
    C. The truck may finish its turn in any one of the two lanes.
    D. Both A and C.

Correct Answer is b

(Long trucks would always swing wide to make a right turn)

Chapter 9
Defensive Driving Techniques

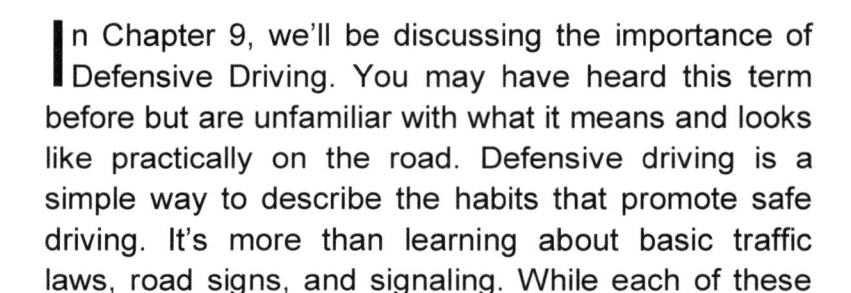

In Chapter 9, we'll be discussing the importance of Defensive Driving. You may have heard this term before but are unfamiliar with what it means and looks like practically on the road. Defensive driving is a simple way to describe the habits that promote safe driving. It's more than learning about basic traffic laws, road signs, and signaling. While each of these are important and deserves your attention, defensive driving encompasses the skills that will reduce your chances of an accident and decrease the risks associated with them.

While learning defensive driving techniques, you may feel like the information is simple or common sense, but these are often the first skills drivers forget once they gain more confidence behind the wheel. It's vital that you continue to practice safe driving habits no matter the amount of experience you have. This chapter will help you learn about speed limits, distractions on the road, merging, switching lanes, and more. You'll also learn about the three-second rule and other tricks and tips you can utilize to stay safe. There are 25 multiple-choice questions covering a wide variety of topics as they relate to defensive driving. You should use these questions as a starting point for learning about safe defensive driving techniques, and then be sure to actively practice them.

Let's begin:

1. When you enter a freeway:

    A. You have the right-of-way.

    B. Give the right-of-way to vehicles that are already driving on the freeway.

    C. Speed up even if you see that there are vehicles on the road ahead.

    D. You must drive faster than the traffic.

Correct Answer is b

(You should be careful when merging into traffic on the freeway. You should signal and enter at the same speed as moving traffic)

2. **If you miss your exit when attempting to get off from a freeway, what should you do?**

    A. Drive to the next exit, and get off the freeway there.
    B. Slow down, get on the shoulder and back up.
    C. Call for assistance.
    D. Take a U-turn from the median of the freeway.

    Correct Answer is a

    (In case you miss your exit, drive to the next exit. Do not stop or reverse your vehicle on the freeway to exit it)

3. **In normal circumstances, what is the ideal following distance between your car and the car that you are following?**

    A. Fifty feet.
    B. The length of your car.
    C. Three seconds behind the vehicle ahead.
    D. Five seconds behind the vehicle ahead.

    Correct Answer is c

    (Using a three-second rule between your car and the car that you are following will provide you the time that you need to spot a hazardous situation and respond quickly)

4. Space cushion is best on which side?

   A. At the front of the vehicle only.
   B. At the right and left sides of the vehicle only.
   C. At the back of your vehicle only.
   D. Each and every side of your vehicle.

Correct Answer is d

(The more space you allow on all sides of your vehicle, the more time you will have to react to hazards on the roadway)

5. When you drive behind a motorcycle, you should:

   A. Provide the following distance of at least one car length.
   B. Provide a distance of at least the length of 5 motorbikes.
   C. Provide at least 4 seconds of distance between your vehicle and the motorcycle.
   D. Provide at least 3 seconds of distance between your vehicle and the motorcycle.

Correct Answer is c

(Usually, the appropriate following distance you can provide is four seconds or more when you drive behind a motorcycle)

6. When you overtake a bicyclist, you must:

   A. Use your four-way flashers.

   B. Drive to the left as much as possible.

   C. Keep your car in the middle of the lane.

   D. Blow your horn to let the bicyclist know.

Correct Answer is b

(When overtaking a bicyclist, lower your speed, provide them as much space as possible and overtake at a safe speed)

7. What causes highway hypnosis?

   A. Looking at the roadway for a very long time.

   B. Getting sleep the day before you drive.

   C. Taking rest stops often.

   D. Getting less sleep the night before your trip.

Correct Answer is a

(Highway hypnosis takes place when you look at the road straight ahead for long periods. Stay attentive regarding your surroundings when you are driving so that you do not fall asleep)

8. What does an orange triangle on the back of a vehicle signify?

A. The vehicle stops frequently.
B. The vehicle is taking wide turns.
C. The vehicle is driven at slower speeds than regular traffic.
D. The vehicle is carrying dangerous materials.

Correct Answer is c

(A reflective orange triangle on the back of a vehicle signifies that the vehicle travels at speeds less than 25 mph)

9. If you see an intersection that has a stop sign, what should you do?

A. Drive when the vehicle in front of you drives.
B. Stop and look, proceed if there's traffic.
C. Don't stop, keep going.
D. Stop and look, proceed if the road is clear.

Correct Answer is d

(If you are stopped at an intersection, you must see look both ways, and in front of you to make sure that the road is clear and safe before you drive ahead)

10. You are making a left hand turn at a multi-lane intersection but traffic from the opposite side is obstructing your view, you must:

A. Increase your speed when you see that the first lane that you need to cross is clear.
B. Wait until you have no obstructions in your view before making the turn.
C. Wait for the crossing guard to let you turn.
D. Block the intersection to make sure you get through.

Correct Answer is b

(Never make a left turn until you have no visual obstructions and can clearly see down the intersection)

11. What is the safe speed at which you can drive your car?

A. Ten miles above the speed limit.
B. Ten miles less than the speed limit.
C. It depends on the climate, road conditions and traffic.
D. It depends on your driving skills.

Correct Answer is c

(If the roads are wet, slippery or under construction, the right speed to drive your car should be less than the posted speed limit)

12. One of the rules of defensive driving is:

A. Look straight ahead as you drive.

B. Stay alert and keep your eyes moving.

C. Expect that other drivers will make up for your errors.

D. Be confident that you can avoid danger at the last minute.

Correct Answer is b

(Driving defensively consists of scanning the road, watching all sides of the road, and checking behind you by using your mirrors)

13. Smoking or lighting up to smoke while you drive:

A. Cannot really be a distraction.

B. Helps to stay awake.

C. Can distract you and cause an accident.

D. Have no effect on the driving abilities.

Correct Answer is c

(Smoking while driving can create dangerous distractions by causing you to take your eyes off the road and your hands off the wheel)

14. How should you merge into traffic onto the freeway?

A. You should drive near the same speed as the other vehicles on the freeway before you merge.
B. 10 mph faster than the traffic on the freeway.
C. At more speed than the stipulated speed limit on the freeway.
D. You should drive at the speed limit posted for the vehicles on the freeway.

Correct Answer is a

(When you enter a freeway, you must enter at a speed that is the same as or near to the other vehicles on the freeway, unless the other vehicles are exceeding the speed limit)

15. While entering a roundabout or traffic circle, drivers:

   A. Have the right of way in case there are two lanes.
   B. Have to yield to drivers that are already in the traffic circle or roundabout.
   C. Have the right-of-way if they enter from the left lane.
   D. Have to stop before they enter.

   Correct Answer is b

   (Drivers who are entering a roundabout or traffic circle must give the right-of-way to drivers who are already inside of the roundabout or traffic circle)

16. To steer clear of any last-minute situations, you have to look down the road where your vehicle will reach in about how much time?

   A. 5 to 15 seconds.
   B. 10 to 15 seconds.
   C. 20 to 25 seconds.
   D. You don't need to estimate it.

Correct Answer is b

(To steer clear of any last-minute situations, you must look down the road 10 to 15 seconds ahead of the vehicle. This is essential if you want to spot any hazards at the earliest)

17. If you are exiting a highway, when should you reduce your speed?

- A. Once you first spot the exit sign on the road.
- B. Once you've come to the red light.
- C. When you move into the exit lane.
- D. When you are on the main road, just before the exit lane.

Correct Answer is c

(You should not reduce your speed while you are on the freeway. Once you are in the exit lane, slow your vehicle down gradually and then maintain the proper exit ramp speed)

18. Which of the following is the right choice if you plan to take a right turn at the corner?

    A. Must take the right turn directly from your lane.
    B. Can just merge into the bicycle lane in case you stop before you turn.
    C. Before making the right turn, you must merge into the bicycle lane.
    D. You do not enter the bicycle lane.

Correct Answer is c

(When making a right turn where there is a bicycle lane, you should merge into the bicycle lane before the corner and then make the turn. Be sure there are no cyclists in your path before merging)

19. In which of these places can you park your car and leave it?

    A. In the middle of a safety zone and the curb.
    B. 10 feet away from a railroad crossing.
    C. Inside of a tunnel.
    D. None of the above.

Correct Answer is d

(You should never park your car and leave in prohibited places like freeways or traffic lanes. You can get a citation and your car towed)

20. If you are driving onto a highway entrance ramp, you can find a gap in freeway traffic by doing which of the following?

   A. By looking in both side-view mirrors only.
   B. By looking in the rearview mirror only.
   C. By looking in the rearview and side-back mirrors.
   D. By looking in both side-view and rearview mirrors and check your blind spot.

Correct Answer is d

(If you drive on a highway entrance ramp and look for a gap, you must look in all the mirrors and check your blind spot)

Defensive Driving Techniques

21. If you want to make a left turn on two-lane, two-way streets or highways, where should you start from?

    A. As Close to the centerline as possible.

    B. From any place in the lane.

    C. From the center of the lane only.

    D. As Close to the outside line as possible.

Correct Answer is a

(If you want to make a left turn on two-lane, two-way streets or highways, you have to stay close to the centerline and make the turn from there)

22. The lane position that you take should _____.

    A. Lower your potential of seeing others and being seen.

    B. Keep your lane safe from other drivers on the road.

    C. Amplify the wind blast that is coming from other vehicles.

    D. All the above.

Correct Answer is b

(If you choose a proper lane position, you can keep yourself out of the blind spot of the other vehicles and this will help you avoid any dangerous situations while on the road)

23. You drive defensively when you:

A. Always put your foot on the brakes.
B. See only the car that is driving in front of you while you are driving.
C. Keep moving your eyes for potential dangers.
D. Keep three car length distances.

Correct Answer is c

(You are driving defensively when you keep your eyes looking down the road for potential hazards)

24. When you drive your vehicle on the road with a truck, you must keep in mind that a truck:

A. Take longer to stop and need more distance than cars.
B. Need less time than cars to pass on an incline.
C. Take less time to spot than cars.
D. Need less time to give a downshift than cars.

Correct Answer is a

(Trucks require longer distances when they have to stop vs. smaller vehicles like cars. This is because trucks are bigger and heavier)

25. What is the significance of minimum speed signs?

A. Allow the traffic to move smoothly.

B. Ensure that the pedestrians are secure.

C. Allow the traffic to move slowly.

D. Check the traffic signal needs in the future.

Correct Answer is a

(Minimum speed limits help enhance the safety on the road and ensure that the traffic can move smoothly)

Chapter 10
The Super Bowl

Welcome to Chapter 10. This chapter is a lengthy review of many of the topics we've discussed so far. This includes lane changes, different types of vehicles, driving under the influence, yielding to pedestrians, and much more. Some of the material may seem like basic or even repetitive information, but it will be the foundation of your driving knowledge and it's important to know all of it. The more you see it, the better you will be at recognizing it. If you find yourself struggling with this chapter, go back to previous chapters and review them until you feel more confident. It's best to take your time and not move forward with new information until you've mastered previous sections.

There are 40 questions in The Super Bowl that you can use to test your overall knowledge of safe driving rules, regulations, habits and techniques. Some are multiple-choice and others are set up in a TRUE/FALSE format. Since this section is a review, you should attempt to answer them instinctually with the first answer that comes to mind. Once you start driving on the road, you'll have much less time to think through possible actions to take. You'll need to make decisions quickly and stick with them so it's important to prepare yourself now before getting behind the wheel of an actual vehicle. Driving is fun but it can also be dangerous. That's we are to help you become a better and safer driver.

Let's Begin:

1. What would you do if you are in a situation where there is an oncoming car to the left and a bicyclist to the right?

    A. Quickly overtake the bike.

    B. Slow down and let both pass.

    C. You will allow the car to pass and then overtake the bike.

    D. Stop at the shoulder.

Correct Answer is c

(You will allow the car to pass and then overtake the bike)

2. **If you want to take a left turn from multi-lane, one-way streets, and highways, where will you start from?**

   A. From the right lane of the street.

   B. From the mid of the intersection.

   C. From the left lane of the street.

   D. Any of the lanes.

   Correct Answer is c

   (From the left lane of the street)

3. **If you are driving and see an emergency vehicle with flashing lights, which of the following will you do?**

   A. You will stop your vehicle immediately.

   B. Continue driving in your lane.

   C. Pull over to the curb and stop.

   D. Reduce the speed and keep driving in your lane.

   Correct Answer is c

   (Pull over to the curb and stop)

4. While driving in traffic, which of the following steps can you take to avoid making any emergency stops?

   A. Drive your vehicle only in the left lane.
   B. Keep a safe following distance from the vehicle ahead of you.
   C. Drive your vehicle slower than the flow of traffic.
   D. Drive your vehicle slower than traffic.

Correct Answer is b

(Keep a safe following distance from the vehicle ahead of you)

5. Who should wear a seat belt while riding in a motor vehicle?

   A. Any passengers under 21 years of age.
   B. Adult passengers in the vehicle only.
   C. The driver of the vehicle.
   D. All of the above.

Correct Answer is d

(All of the above)

6. **When entering the interstate on a short entrance ramp where there is no acceleration lane, you should:**

    A. Quickly enter the far-left lane on the Interstate and increase the speed.
    B. Make use of the emergency lane and increase speed.
    C. Increase your speed, enter the interstate and find an open space in the traffic.
    D. None of the above.

    Correct Answer is c

    (Increase your speed, enter the interstate and find an open space in the traffic)

7. **When you drive on the road and hear a siren coming from behind you, what will you do?**

    A. Increase your speed and get out of the way.
    B. Pull over to the right side and then stop.
    C. Stop immediately.
    D. Reduce the speed of your vehicle only.

Correct Answer is b

(Pull over to the right side and then stop)

8. You've reached a four-way stop sign, you have to give the right-of-way to all motorists who have reached the four-way before you.

   A. TRUE
   B. FALSE

Correct Answer is a

(True)

9. You are driving your vehicle on the highway, and traffic is merging into your lane. Which of these should you do?

   A. Provide space to the merging traffic if there is a possibility.
   B. Keep your position as is.
   C. Blow your horn to tell the driver not to merge.
   D. Signal your right-of-way by increasing the speed of your vehicle.

Correct Answer is a

(Provide space to the merging traffic if there is a possibility)

10. When you are switching lanes while driving, your blind spot can be seen by using:

    A. The outside mirrors.

    B. Your eyes, turning your head and looking over your shoulders.

    C. The rearview mirror.

    D. Both A and C.

Correct Answer is b

(Your eyes, turning your head and looking over your shoulders)

11. If you are planning to enter a freeway where there's traffic, how must you enter?

    A. Decrease your speed and merge.

    B. First, check behind you, and then check the vehicle in front of you.

    C. Check the vehicles that are behind the gap only.

    D. Stop your vehicle, check whether there's available gap space so that your vehicle can enter.

Correct Answer is b

(First, check behind you, and then check the vehicle in front of you)

12. What is the minimum blood alcohol concentration (BAC), which is illegal for a person 21 years of age or older to drive?

    A. 3%

    B. 8%

    C. 5%

    D. Any concentration is illegal.

Correct Answer is b

(8%)

13. If you stop for a school bus that is dropping children off from school, you must:

    A. Speed up quickly.

    B. Watch for children walking across the road.

    C. Drive if you don't see any children getting off.

    D. Turn around.

Correct Answer is b

(Watch for children walking across the road)

14. While taking a left at an intersection, which of the following should be adhered to?

A. You must always give the right-of-way to pedestrians and oncoming traffic.
B. You do not have to give the right-of-way to pedestrians and oncoming traffic.
C. Oncoming vehicles and pedestrians should give way.
D. None of the above.

Correct Answer is a

(You must always give the right-of-way to pedestrians and oncoming traffic)

15. If you are following a driver who seems distracted, you should:

A. Increase the distance behind the driver ahead.
B. Pay them no attention.
C. Keep the same following distance.
D. Decrease the following distance between your vehicle and the vehicle in front

Correct Answer is a

(Increase the distance behind the driver ahead)

16. If you enter a school zone during school hours, what is the speed limit?

    A. 5 mph

    B. 15 mph

    C. 20 mph

    D. 10 mph

Correct Answer is b

(15 mph)

17. When changing lanes, keep in mind that you must:

    A. Turn your head and check your blind spot.

    B. Initiate your turning signal to alert others that you are changing lanes.

    C. Ensure that the lane that you are entering is clear.

    D. All of the above.

Correct Answer is d

(All of the above)

18. How can you avoid hydroplaning on the road?

    A. By increasing speed and driving more quickly.

    B. By driving through deeper water.

    C. By decreasing speed and driving more slowly.

    D. By driving through puddles of water.

Correct Answer is c

(By decreasing speed and driving more slowly)

19. Which is the proper way to use a highway exit ramp?

    A. Overtake the slower traffic in the exit ramp.

    B. Reduce the speed once you move onto the exit ramp.

    C. Maintain your same speed when exiting the ramp.

    D. Reduce your speed before you enter the exit ramp.

Correct Answer is b

(Reduce the speed once you move onto the exit ramp)

20. Which of the following should you do in bad weather?

    A. Drive your vehicle in high gear.

    B. Drive your car in 1st gear.

    C. Steer and apply your brakes smoothly.

    D. Steer the vehicle off the road.

Correct Answer is c

(Steer and apply your brakes smoothly)

21. If a passenger vehicle signals to re-enter the main road following a stop, you must:

    A. Increase your speed and pass the vehicle.

    B. Move to the right lane of the vehicle.

    C. Give way to the passenger vehicle.

    D. Blow your horn to make them aware of your presence.

Correct Answer is c

(Give way to the passenger vehicle)

22. You should treat a railroad crossbuck sign similar to the yield sign.

    A. True
    B. False

    Correct Answer is a

    (True)

23. If you want to make a quick turn, how will you position your hands?

    A. Position your hands on the opposite sides of the steering wheel.
    B. Position both of your hands on the same side of the steering wheel.
    C. Position one hand on the top and the other hand on the bottom of the steering wheel.
    D. Position your hands next to each other on the steering wheel.

    Correct Answer is a

    (Position your hands on the opposite sides of the steering wheel)

24. In the morning or night, rain or snow, seeing and being seen can be difficult. Which of the following is a great way to ensure that the other drivers know that you are also on the road?

   A. Honk your horn.
   B. Turn your parking lights on and off.
   C. Turn your headlights on.
   D. Turn on your emergency flashlight.

Correct Answer is c

(Turn your headlights on)

25. If you are driving in foggy weather, it is best to drive using:

   A. Your yellow headlights.
   B. Your headlights on a low beam.
   C. With the four-way flashers.
   D. Your headlights on a high beam.

Correct Answer is b

(Your headlights on a low beam)

26. If you are about to be rear-ended, which of the following you must not do?

- A. Take off your seat belt.
- B. Apply your brakes.
- C. Brace yourself for the impact.
- D. All of the above.

Correct Answer is a

(Take off your seat belt)

27. An octagonal sign signifies a:

- A. Speed limit sign.
- B. Stop sign.
- C. Hospital sign.
- D. Railroad warning sign.

Correct Answer is b

(Stop sign)

28. **If the traffic light turns green but a lot of traffic obstructs the intersection ahead, what should you do?**

    A. Drive ahead as the traffic light is green.
    B. Wait for traffic to clear before entering the intersection.
    C. Wait for the red light before entering the intersection.
    D. Take a right turn onto the bicycle land and check if there is another route.

    Correct Answer is b
    (Wait for traffic to clear before entering the intersection)

29. **If a vehicle has just stopped and parked on the side of the street, what should you do when driving past that vehicle?**

    A. Consider that the driver may swing its door open while passing.
    B. Blow your horn.
    C. Pull up in front the parked vehicle.
    D. It's an emergency, call 911.

Correct Answer is a

(Consider that the driver may swing its door open while passing)

30. The adjustment of a seat belt should be made so that it is:

A. Slacked enough for comfort.

B. Snugged enough for comfort.

C. Tight enough for less movement.

D. Secures your lower abdomen only.

Correct Answer is b

(Snugged enough for comfort)

31. If you are driving your vehicle on major highways, you must:

A. Stay ready to respond to any dangers on the road.

B. Stay aware of your surroundings

C. Move your eyes constantly.

D. All of the above.

Correct Answer is d

(All of the above)

32. Following directions given by a flagger in a construction zone is not necessary.

    A. True

    B. False

Correct Answer is b
(False)

33. Which situation is legal to park your vehicle in a handicap parking spot?

    A. Not in any circumstance.

    B. When you want to park your car for a minute or two.

    C. When unloading your vehicle.

    D. When all parking spots in the parking lot are full.

Correct Answer is a
(Not in any circumstance)

34. You are driving your vehicle in a construction zone, you should:

   A. Not drive into a construction zone.

   B. Slow down.

   C. Pull over.

   D. Speed up slightly.

Correct Answer is b

(Slow down)

35. If the tire of your vehicle blows out, what should you do?

   A. Give it more gas to have more control.

   B. Hold the steering wheel firmly, keep calm while applying a light break.

   C. Pull to the side as quickly as possible.

   D. Break hard and fast.

Correct Answer is b

(Hold the steering wheel firmly, keep calm while applying a light break)

36. Which of the following describes a blind spot?

   A. Areas of the road that only the passenger can see.
   B. The spot that cannot be seen without turning your head and looking over your shoulders.
   C. The area that's near the crosswalk.
   D. The spots that cloud your vision when you are tired.

Correct Answer is b

(The spot that cannot be seen without turning your head and looking over your shoulders)

37. You can only Cross solid yellow lines when:

   A. On the weekends only.
   B. When its night hours.
   C. It's between 8am and 5pm.
   D. When you're making turns onto another street.

Correct Answer is d

(When you're making turns onto another street)

38. Which of these should you do while you drive at night?

A. Use your parking lights.
B. Cut off the lights on your vehicle so you don't blind oncoming traffic.
C. Increase the distance from the vehicle ahead.
D. Decrease the distance from the vehicle ahead.

Correct Answer is c

(Increase the distance from the vehicle ahead)

39. Which of the following should a driver do?

A. Focus on the right side of the road only.
B. Focus on the vehicle to the rear only.
C. Focus on the right, left, front and rear of your vehicle.
D. Both B and C.

Correct Answer is c

(Focus on the right, left, front and rear of your vehicle)

40. If a person fails to report an accident that led to property damage of $1000 or more, it will not be charged as a criminal offense.

   A. True
   B. False

   Correct Answer is b
   (False)

Chapter 11
Bonus Cheat Sheet I

Chapter 11 has been added as a bonus chapter that will serve as a cheat sheet for the most seen and recognizable questions on the driver's license exam. Cheat Sheets doesn't mean you are cheating. It simply means these questions are repetitive and will give you more of an insight of what's most likely to be on the exam. There are 28 situational questions in this section that go into detail about how to handle various occurrences on the road. While no one can possibly prepare for every circumstance that may arise while driving, these situations are a great starting point. Read each question, then memorize the answer. This way, you'll have no problem with these questions doing the exam.

The questions in Chapter 11 are purposefully no multiple choice and random because every time you drive will be unique. You'll need to react suddenly in many instances, and you cannot afford to take the wrong action or make an incorrect decision. Doing so could cause an accident and could even prove to be fatal for yourself or others. The most important thing to remember as you're reviewing this material and learning to drive is to expect the unexpected. Make sure you are practicing defensive driving at every opportunity to mitigate the risks of driving. As you gain confidence behind the wheel, continue to display the safe driving habits which you've learned throughout this course.

Let's begin:

1. What do you do if you're about to approach a railroad crossing that isn't equipped with traffic-directing signals?

   Reduce your speed and expect to stop.

   "Reduce your speed and expect to stop and wait before approaching an unmarked railroad crossing. Be sure there are no trains coming from any direction on either track before crossing."

2. You can see that the left arm and hand of the driver are extended downward. What does the driver intend to do according to this hand gesture?

   Stop.

"When a driver's left arm and hand are extended downward, that means they're about to stop."

3. When do you have the right-of-way?

When you are within a traffic circle already.

"When entering a traffic circle or roundabout, drivers must give the right-of-way to those already inside."

4. A car behind you starts to overtake you. What should you do?

Stay in your own lane after reducing your speed slightly.

"Decrease your speed slowly and stay to the right if another vehicle passes you on the left."

5. A "No stopping" sign indicates that you can only stop if told to do so by a police officer. When else can you stop?

You may also stop to prevent conflicts with other vehicles.

"When you see a "No Stopping" sign, that means you should either stop to follow a traffic light, obey a police officer, or avoid a conflict with other traffic."

6. What should you do before you leave a parallel-curb parking spot?

Turn your head and check for traffic.

"Turn your head and check over your shoulder before exiting a parking spot to drive back into traffic."

7. How can you avoid tiredness on long trips?

By stopping at frequent intervals for a break.

"You should arrange frequent breaks to stop and rest during long trips to avoid tiredness."

8. You see a stop sign and a crosswalk present at an intersection, but there is no stop line. You must:

Stop your vehicle before the crosswalk.

"If there is a stop line at a stop sign, you must stop before you reach it."

9. A driver consumes non-prescription drugs. What should the driver do before driving?

The driver must read the drug's labels.

"Many over-the-counter prescriptions have the potential to impair the ability to drive safely. Check the label for warnings about the consequences of any drugs you take"

10. You are pulling out of your garage. You notice that children are playing close by. What should you do?

   To make sure the path is clear, look out the back windshield and walk to the rear of the vehicle.

   "Look out the back windshield for pedestrians and other barriers while backing up a vehicle."

11. You are exiting a freeway. What should you do?

   Check the speed limit for the lane you're about to enter.

   "It's important to be aware of other vehicles passing while approaching the freeway. Check the speedometer of your vehicle after looking at the latest posted speed limit."

12. Your vehicle can be no more than _____ from the curb while parallel parked.

    One foot.

    "Your vehicle can be no more than one foot from the curb while parallel parked."

13. In a construction area, fines for moving traffic violations are ____ than otherwise.

    Two times

    "In highway construction or maintenance areas where employees are involved, fines for moving traffic offences are two times the normal fine."

14. On a road, you notice the road-line turning inward towards you. What does it mean?

    It means that the road ahead is becoming narrower.

"On a road, when you see the road-line starts to slant inward, it means the road ahead is narrowing."

15. You are driving behind another vehicle under normal driving conditions. Which rule should you follow?

    You should follow the three-second rule.

    "You should preserve a following distance of at least three seconds when under normal driving conditions."

16. You are braking on a wet and slippery surface. The breaks should be applied:

    Earlier than normal.

    "To prevent locking the brakes while braking on wet and slippery surfaces, such as roads coated with rain or snow, apply the brakes early in a slow and steady manner."

17. When are you allowed to drive under the influence of any drug or medication that hinders your ability to drive safely?

   Never.

   "Driving under the influence of any drug that hinders your ability to drive safely is prohibited. Alcohol, over-the-counter or prescription medications all fall under this category."

18. You see a solid yellow line next to a broken yellow line. This indicates that the vehicles:

   Near the broken line are permitted to pass.

   "Oncoming traffic lanes are separated by yellow lines. You can pass if there is a broken yellow line next to your driving lane."

19. You see a pedestrian crossing but there is no crosswalk. What should you do?

   Stop, wait and let the pedestrian cross.

"Even if there isn't a designated crosswalk, drivers must give way to pedestrians crossing the street."

20. If you want to see if any vehicles are in your blind spots, you should:

Turn your head and look over your shoulders.

"Blind spots are on the rear left and right side that you can't see through your mirrors. Turn your head and look over your shoulder to check and see if any vehicles are in your blind spot."

21. What should you do when traffic is merging into your lane while driving on the expressway?

Make room and let merging traffic in.

"Where it's possible, you should leave space for vehicles to merge into your lane as traffic permits."

22. How should you be driving if you are about to merge onto the freeway?

You should be driving at or near the speed of the freeway traffic.

"If you are merging onto the freeway, approach the freeway at or near the speed of traffic."

23. When should you drive across an intersection if you know you'll be blocking the intersection as the light turns red?

Never.

"You must not approach an intersection until you can cross completely before the light turns red, even though the signal is orange. You will be cited if you block the intersection"

24. What should you do when you see a traffic signal with a red arrow pointing to the right?

Wait until the light turns green before turning in that direction.

"A red arrow is treated the same as a red light. Remain stopped until a green light or a green arrow appears."

25. You hear an emergency vehicle and see flashing lights behind you. You must yield the right-of-way by:

    Driving to the right side of the road as possible and stopping.

    "You are required to yield to any emergency vehicle when it is using its flashing lights and siren."

26. When is it permissible to lawfully travel around or under a railroad crossing gate?

    Under no circumstances.

"At a railroad crossing, do not go under or around any lowered gate. Once the gate is opened, do not cross the tracks until you can see in all directions and are certain that no trains are approaching."

27. What should you do if you begin to lose traction due to water on the road?

Decrease your speed without slamming on the brakes.

"This is known as "hydroplaning." Decelerate cautiously and do not apply the brakes if the car begins to hydroplane."

28. When driving into an environment where children are playing, you should expect the children to:

Run in front of your vehicle without looking.

"When children are around, you should always slow down and exercise extra caution."

Chapter 12
Bonus Cheat Sheet II

Similar to Chapter 11, Chapter 12 has also been created to provide extra assistance in the form of another cheat sheet. There are 28 questions in this section; some are fill-in-the-blank and others are short response questions. Each question comes with a description of why the correct answer is the best and safest answer. At times, it may seem like more than one action could be taken safely in a situation; therefore, you might feel like there could be more than one correct answer. However, keep in mind this cheat sheet is only a starting point for the situations that you will face out on the road. Some answers are more definitive than others. For example, you should never drive under the influence. You also should not drive if you're tired. Try to focus most on the questions that have more clear-cut answers, and then think through the questions that have longer explanations on your own. Thinking through them now will give you an advantage should you face these situations in real life later on.

You will still need to gain real-world experience driving out on the road, as that will be what prepares you most for the unexpected. This cheat sheet does not contain every situation you will face behind the wheel. You'll need to learn how your vehicle moves and operates under different road conditions (such as heavy rainfall) and experience situations that force you to respond quickly. Hopefully, this cheat sheet will be helpful to you as you prepare to start driving.

Let's begin:

1. In the event that your turn signals fail, you must use _____ to show that you are turning.

   Hand signals.

   "Use hand signals to show that you are about to make a turn if your vehicle's turn signals do not function."

2. You are driving and you start to experience fatigue. What is the best thing you can do?

   Pull over and stop driving.

"Driving when fatigued is as dangerous as driving under the influence of alcohol. The easiest thing to do if you start to feel exhausted when driving is to pull over."

3. You are in traffic and another vehicle cuts you off. What should you do?

Just let it go and don't retaliate.

"Never take it personally if another vehicle cuts you off in traffic. This can help you to prevent the risks of road rage"

4. Where should you keep a space cushion?

Keep a space cushion on all the sides of the vehicle.

"It is safer to keep a cushion of space on all side of the vehicle to ensure that you'll have time to respond to dangers on the roadway."

5. It is important to _____ your speed while making a turn.

Decrease.

"You must slow down by decreasing your speed when making a turn to maintain and control your vehicle."

6. What should you do while driving on slick roads?

Make turns with a lower speed

"Driving too fast on slippery roads is often risky. This is particularly true when you make turns and drive through curves."

7. When do roads freeze quickly?

Shaded.

"In cold or rainy season, take extra precautions on stretches of road that are shaded by trees or buildings."

8. You come across an intersection with a yield sign. What should you do?

   Decrease your speed and yield the right-of-way to other vehicles.

   "You must slow down and yield the right-of-way to traffic in the intersection or lane you are approaching if you see a yield sign."

9. You are parking uphill next to a curb. What should you do after setting the parking brake?

   Make sure that you turn and keep the steering wheel away from the curb.

   "Set your parking brake and turn the steering wheel away from the curb while parking facing uphill on a street with a curb."

10. If you are driving in fog, rain or snow, what should you use?

    Headlights with low beams.

"While driving in snow, rain, or fog, low beam headlights should be used in these conditions."

11. What should you do when you have a tire blowout?

Slow down gradually and use your brakes lightly.

"If you have a tire blowout, slowly reduce your speed by lifting your foot off the accelerator."

12. If your vehicle starts to skid. What should you do?

Don't apply the brakes.

"Do not use the brakes if the vehicle starts to skid. Braking can worsen the skid."

13. When it's raining, you should be extra cautious when making turns. Especially during:

The first half hour of rain.

"During the first half hour of rain, be extra cautious when making turns. The roads are slick with oil from vehicle traffic."

14. When two drivers reach a four-way intersection at the same time:

The right-of-way belongs to the driver on the right.

"If you both arrive at a four-way junction at the same time, yield the right-of-way to the driver on your right. After the vehicle passes, you can proceed."

15. What are the orange-colored signs?

Orange colored signs are work zone signs.

"Work zone signs are usually diamond-shaped, orange, and have black letters or markings, and they act as an alert that people are located near or on the highway."

16. What does a pennant-shaped sign mean?

   This is a no passing zone.

   "No-passing zones are normally bright yellow pennant-shaped signs. The sign would also have words confirming that it is a no-passing zone."

17. Regulatory signals are normally of _____ and must be followed at all times.

   The color white.

   "The flow of traffic is controlled by regulatory signs. Motorists should always follow them and they are normally white with black markings."

18. What is indicated by a pentagonal sign?

   School zones use pentagonal signs.

   "You are driving in a school crossing zone if you see a pentagonal shaped signs."

19. What does a diamond-shaped sign mean?

   Warns you of any hazards.

   "Diamond-shaped signs alert drivers of probable or current road dangers. These signs are normally yellow or orange in color."

20. You are approaching a traffic signal with flashing yellow light. What should you do?

   Decrease your speed and proceed with caution.

   "A hazard is indicated by a flashing yellow light. You should decrease your speed and move ahead cautiously."

21. What should drivers do when an officer is directing traffic at a traffic light?

   Follow the officer's directions.

"Even if the traffic signal is working, you must obey the officer directing traffic."

22. If a driver enters public traffic from a private lane or a driveway, the driver should:

Yield the right-of-way to the drivers who are already on the public road.

"If you're approaching traffic from a driveway or a private lane, you must yield to oncoming traffic. When you're ready, merge safely into traffic."

23. When two drivers reach an intersection at the same time:

The driver on the left should yield the reight-of-way to the driver on the right.

"You must yield the right-of-way to a vehicle to your right if you arrive at an open intersection at the same time. When it is safe to do so, you can proceed."

24. You're driving on the interstate and you've passed your exit. What should you do?

   Don't slam on your brakes, get off on the next exit.

   "Don't slam on your brakes. Proceed to the next exit if you miss your exit."

25. What should drivers do while they approach a blinking red traffic light?

   Treat the light as if it were a stop sign.

   "Blinking red lights are just like stop signs."

26. What should you do if traffic is being directed by a crossing guard in a school zone?

   Pay attention to the crossing guard's instructions and follow them.

"Be ready to slow down and exercise caution when a sign, crossing guard, or law enforcement officer directs you to do so. In a school district, drive with extra caution."

27. What is indicated by a sign that says "end school zone"?

    The end of a speed limit zone.

    "During indicated times, never surpass the school zone speed limit."

28. What is indicated by a crossbuck sign?

    A railroad crossing is indicated by a crossbuck signs.

    "At a railroad crossing, crossbuck signals instruct drivers to yield to trains. Drivers must never to try to outrun a train."

Chapter 13
Preparing for Your Exam

It's time to prepare for your exam. You've made it through the chapters and you're ready to get started. Before you can take your exam, you must first understand what you need to do to be successful.

Let's first review some strategies for success:

- **Study:**

In order to be successful on your exam, you must take advantage of the resources provided in this book and study very closely. The more you study, the easier it will become to differentiate between rules, signs, and signals. You must ensure you're not simply memorizing the content but understanding it. The information you learn in these materials will be crucial throughout the entirety of your life as a driving adult.

- **Create a Study Plan:**

For many people, simply studying and reading materials is not enough. Sometimes it's best to set a very specific schedule and study plan. Plan out how many hours per day you will study and what your goals will be during each study session. Make every effort to stick to your plans to ensure your success. Most importantly, reward your efforts to remain motivated.

- **Take plenty of Practice Tests:**

We all know the saying, "practice makes perfect". This phrase applies to taking your driver's or learner's permit exam as well. You must take tons of practice tests to get familiar with the style of questions asked and ensure you're properly prepared for your exam.

- **Get lots of Rest:**

It's easy to force yourself to stay up all night studying in preparation for your exam, however, this must be avoided. The night before your exam should consist of a short preparation session and then lots of sleep. You need to ensure you're well-rested and prepared for the following day.

- **Stay Positive:**

While it may seem daunting and overwhelming at times, you are the only one that can achieve your goals. You must stay optimistic and dedicate yourself to learning, practicing, and preparing for your exam. Visualize your success and gain confidence through preparation.

Chapter 14
Applying for Your Driver's License or Permit

Now that you've read through the success principles and ran through several practice questions, it's time to obtain your driver's license or permit. In the following sections, we will explore the processes and requirements for each of these actions.

A driver's license is an official authorization to operate a motor vehicle. It is mandatory throughout the U.S. that all individuals behind the wheel maintain an active and valid driver's license or permit.

Obtaining your Driver's License:

- Fill out a Driver's License Application
- Collect required documents and materials (proof of identification, U.S. citizenship or lawful presence, social security number, residency, proof of insurance)

- Schedule an appointment at a nearby motor vehicle office or division
- Pass a vision exam
- Pass the behind-the-wheel test
- Pay application fee

A learner's permit is very similar to a driver's license, but it only provides limited driving capabilities. Restrictions may limit the hours an individual can drive and restrict one's ability to drive without an adult present.

Obtaining your Learner's Permit:

- Complete a driver's education course (online or in-person)
- Fill out a Learner's Permit Application
- Collect required documents and materials (proof of identification, U.S. citizenship or lawful presence, residency, social security number, course completion, school attendance verification)
- Schedule an appointment at a nearby motor vehicle office or division
- Pass a vision exam
- Pass the online permit test
- Pay application fee

Remember that requirements may change slightly depending on the time that you've applied. Be sure to validate requirements against your state's outlined resources to ensure you're best prepared. There may be changes and differences in course types and hours, documentation requirements, age restrictions, and time limits.

Conclusion

You've made it to the end of this course! Throughout this course, you've had the opportunity to take part in a variety of lessons in preparation for your driver's exam.

In chapter one, you learned about the various laws and regulations surrounding the operation of vehicles while under the age of 21. In chapter two, you learned about safety conditions and rules of engagement on the road. In chapters three and four, you obtained insight into the different road signs, traffic lights, and signals to ensure you can easily understand different road conditions and situations. In chapter five, you learned how to safely conduct turns and maneuver your vehicle. Chapter six served as a review of road signs and chapter seven took you through the process of identifying and understanding the risks associated with driving under the influence.

In chapter eight, we discussed habits for success when driving. Chapter nine took you through defensive driving. An overview of the previous sections was provided in chapter ten. Chapters eleven and twelve were bonus chapters that provided cheat sheets to aid your learning. In chapter thirteen, you learned some key strategies for success for your exam. In chapter fourteen, we've identified the requirements for obtaining either a driver's license or permit.

To ensure you are best prepared, it is recommended that you review the contents of these materials several times. The most successful students have reviewed these materials at least two to three times and are confident in their answers.

It's time to schedule your exam and get ready to obtain your license or permit.

Until then, Good Luck my friend and we would Love to see a Picture of you holding your brand new Certificate, Driver's License or Learner's Permit very Soon!

Would You Leave Us A Review?

My staff and I work tirelessly to ensure that we provide a holistic approach to educating and preparing our readers.

We welcome ALL Feedback both Positive and Negative, even if it's just one sentence, this will help us further enhance our training and provide more value.

We would definitely Appreciate it either way!

Review.Vast-Pass.com

-Thank You

Stanley Vast

Certificate Of Achievement

Congratulations! You Did It!

Go to Cert.Vast-Pass.com to receive your Certificate.

Simply enter your **Name** and **Date** and that's it!

New Jersey Driver's Practice Tests

Subscribe to Get Your Private Mobile App + Digital Flashcards

VastPass

Go to **NJ.Vast-Pass.com** to get your

Private Mobile App & Digital Flashcards!

Resources Page

Mothers Against Drunk Driving (MADD) – Support MADD, too many of our love ones have been lost due to Driving Under The Influence. Donate to Stop Drunk Driving and Save a Life Today!
https://www.Madd.org/

Donations

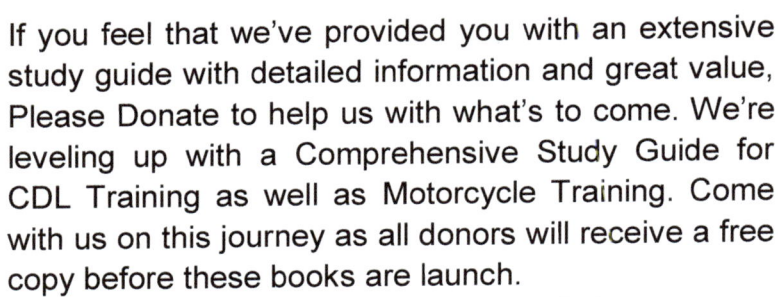

If you feel that we've provided you with an extensive study guide with detailed information and great value, Please Donate to help us with what's to come. We're leveling up with a Comprehensive Study Guide for CDL Training as well as Motorcycle Training. Come with us on this journey as all donors will receive a free copy before these books are launch.
Donate.Vast-Pass.com Spear Market Sales

Thank you for your generosity.